Lecture Notes in Physics

For information about Vols. 1–44, please contact your bookseller or Springer-Verlag.

Vol. 45: Dynamical Concepts on Scaling Violation and the New Resonances in e^+e^- Annihilation. Edited by B. Humpert. VII, 248 pages. 1976.

Vol. 46: E. J. Flaherty, Hermitian and Kählerian Geometry in Relativity. VIII, 365 pages. 1976.

Vol. 47: Padé Approximants Method and Its Applications to Mechanics. Edited by H. Cabannes. XV, 267 pages. 1976.

Vol. 48: Interplanetary Dust and Zodiacal Light. Proceedings 1975. Edited by H. Elsässer and H. Fechtig. XII, 496 pages. 1976.

Vol. 49: W. G. Harter and C. W. Patterson, A Unitary Calculus for Electronic Orbitals. XII, 144 pages. 1976.

Vol. 50: Group Theoretical Methods in Physics. 4th International Colloquium. Nijmegen 1975. Edited by A. Janner, T. Janssen, and M. Boon. XIII, 629 pages. 1976.

Vol. 51: W. Nörenberg and H. A. Weidenmüller. Introduction of the Theory of Heavy-Ion Collisions. 2nd enlarged edition. IX, 334 pages. 1980.

Vol. 52: M. Mladjenović, Development of Magnetic β-Ray Spectroscopy. X, 282 pages. 1976.

Vol. 53: D. J. Simms and N. M. J. Woodhouse, Lectures on Geometric Quantization. V, 166 pages. 1976.

Vol. 54: Critical Phenomena. Sitges International School on Statistical Mechanics, June 1976. Edited by J. Brey and R. B. Jones. XI, 383 pages. 1976.

Vol. 55: Nuclear Optical Model Potential. Proceedings 1976. Edited by S. Boffi and G. Passatore. VI, 221 pages. 1976.

Vol. 56: Current Induced Reactions. International Summer Institute, Hamburg 1975. Edited by J. G. Körner, G. Kramer, and D. Schildknecht. V, 553 pages. 1976.

Vol. 57: Physics of Highly Excited States in Solids. Proceedings 1975. Edited by M. Ueta and Y. Nishina. IX, 391 pages. 1976.

Vol. 58: Computing Methods in Applied Sciences. Proceedings 1975. Edited by R. Glowinski and J. L. Lions. VIII, 593 pages. 1976.

Vol. 59: Proceedings of the Fifth International Conference on Numerical Methods in Fluid Dynamics. 1976. Edited by A. I. van de Vooren and P. J. Zandbergen. VII, 459 pages. 1976.

Vol. 60: C. Gruber, A. Hintermann, and D. Merlini, Group Analysis of Classical Lattice Systems. XIV, 326 pages. 1977.

Vol. 61: International School on Electro and Photonuclear Reactions I. Edited by C. Schaerf. VIII, 650 pages. 1977.

Vol. 62: International School on Electro and Photonuclear Reactions II. Edited by C. Schaerf. VIII, 301 pages. 1977.

Vol. 63: V. K. Dobrev et al., Harmonic Analysis on the n-Dimensional Lorentz Group and Its Application to Conformal Quantum Field Theory. X, 280 pages. 1977.

Vol. 64: Waves on Water of Variable Depth. Edited by D. G. Provis and R. Radok. 231 pages. 1977.

Vol. 65: Organic Conductors and Semiconductors. Proceedings 1976. Edited by L. Pál, G. Grüner, A. Jánossy and J. Sólyom. 654 pages. 1977.

Vol. 66: A. H. Völkel, Fields, Particles and Currents. VI, 354 pages. 1977.

Vol. 67: W. Drechsler and M. E. Mayer, Fiber Bundle Techniques in Gauge Theories. X, 248 pages. 1977.

Vol. 68: Y. V. Venkatesh, Energy Methods in Time-Varying System Stability and Instability Analyses. XII, 256 pages. 1977.

Vol. 69: K. Rohlfs, Lectures on Density Wave Theory. VI, 184 pages. 1977.

Vol. 70: Wave Propagation and Underwater Acoustics. Edited by J. Keller and J. Papadakis. VIII. 287 pages. 1977.

Vol. 71: Problems of Stellar Convection. Proceedings 1976. Edited by E. A. Spiegel and J. P. Zahn. VIII, 363 pages. 1977.

Vol. 72: Les instabilités hydrodynamiques en convection libre forcée et mixte. Edité par J. C. Legros et J. K. Platten. X, 202 pages. 1978.

Vol. 73: Invariant Wave Equations. Proceedings 1977. Edited by G. Velo and A. S. Wightman. VI, 416 pages. 1978.

Vol. 74: P. Collet and J.-P. Eckmann, A Renormalization Group Analysis of the Hierarchical Model in Statistical Mechanics. IV, 199 pages. 1978.

Vol. 75: Structure and Mechanisms of Turbulence I. Proceedinas 1977. Edited by H. Fiedler. XX, 295 pages. 1978.

Vol. 76: Structure and Mechanisms of Turbulence II. Proceedings 1977. Edited by H. Fiedler. XX, 406 pages. 1978.

Vol. 77: Topics in Quantum Field Theory and Gauge Theories. Proceedings, Salamanca 1977. Edited by J. A. de Azcárraga. X, 378 pages 1978.

Vol. 78: Böhm, The Rigged Hilbert Space and Quantum Mechanics. IX, 70 pages. 1978.

Vol. 79: Group Theoretical Methods in Physics. Proceedings, 1977. Edited by P. Kramer and A. Rieckers. XVIII, 546 pages. 1978.

Vol. 80: Mathematical Problems in Theoretical Physics. Proceedings, 1977. Edited by G. Dell'Antonio, S. Doplicher and G. Jona-Lasinio. VI, 438 pages. 1978.

Vol. 81: MacGregor, The Nature of the Elementary Particle. XXII, 482 pages. 1978.

Vol. 82: Few Body Systems and Nuclear Forces I. Proceedings, 1978. Edited by H. Zingl, M. Haftel and H. Zankel. XIX, 442 pages. 1978.

Vol. 83: Experimental Methods in Heavy Ion Physics. Edited by K. Bethge. V, 251 pages. 1978.

Vol. 84: Stochastic Processes in Nonequilibrium Systems, Proceedings, 1978. Edited by L. Garrido, P. Seglar and P. J. Shepherd. XI, 355 pages. 1978

Vol. 85: Applied Inverse Problems. Edited by P. C. Sabatier. V, 425 pages. 1978.

Vol. 86: Few Body Systems and Electromagnetic Interaction. Proceedings 1978. Edited by C. Ciofi degli Atti and E. De Sanctis. VI, 352 pages. 1978.

Vol. 87: Few Body Systems and Nuclear Forces II, Proceedings, 1978. Edited by H. Zingl, M. Haftel, and H. Zankel. X, 545 pages. 1978.

Vol. 88: K. Hutter and A. A. F. van de Ven, Field Matter Interactions in Thermoelastic Solids. VIII, 231 pages. 1978.

Vol. 89: Microscopic Optical Potentials, Proceedings, 1978. Edited by H. V. von Geramb. XI, 481 pages. 1979.

Vol. 90: Sixth International Conference on Numerical Methods in Fluid Dynamics. Proceedings, 1978. Edited by H. Cabannes, M. Holt and V. Rusanov. VIII, 620 pages. 1979.

Lecture Notes in Physics

Edited by J. Ehlers, München, K. Hepp, Zürich
R. Kippenhahn, München, H. A. Weidenmüller, Heidelberg
and J. Zittartz, Köln
Managing Editor: W. Beiglböck, Heidelberg

121

Frederik W. Wiegel

Fluid Flow Through Porous Macromolecular Systems

Springer-Verlag
Berlin Heidelberg GmbH 1980

Author

Frederik W. Wiegel
Department of Applied Physics
Twente University of Technology
Enschede
The Netherlands

ISBN 978-3-540-09973-4 ISBN 978-3-540-39203-3 (eBook)
DOI 10.1007/978-3-540-39203-3

Library of Congress Cataloging in Publication Data. Wiegel, Frederik W. 1938- Fluid flow through porous macromolecular systems. (Lecture notes in physics; 121) Bibliography: p. Includes index. 1. Fluid dynamics. 2. Diffusion. 3. Cell membranes. 4. Statistical mechanics. I. Title. II. Title: Porous macomolecular systems. III. Series. [DNLM: 1. Biological transport. 2. Body fluids--Metabolism. 3. Macromolecular systems. QU105 W645f] QC155.W53. 574.19'1. 80-13872

Originally published by Springer-Verlag Berlin Heidelberg New York in 1980

2153/3140-543210

CONTENTS

Appendix: Comments on entanglements and the excluded volume problem

PREFACE

This short monograph is based on lectures which I presented at the Los Alamos Scientific Laboratory, the National Bureau of Standards and the sixth Oaxtepec symposium on statistical mechanics.

I have tried to keep the material as self-contained as possible. The contents can be divided into three parts: part A consists of sections 1 - 7, part B of sections 8 - 18, and part C of sections 19 - 28. These three parts have been written in such a way that each of them forms an independent unit. Part A will appeal to readers primarily interested in statistical thermodynamics, part B to chemical physicists, and part C to biophysicists. Although most of the text is devoted to the development of the theory, I have made an effort to compare some of the predicted phenomena with experimental data.

The models which form the basis of the following calculations are phenomenological. The material in this booklet, therefore, will complement the a priori calculations which form the subject of two review papers in print: "Theories of Lipid Monolayers", by F.W. Wiegel and A.J. Kox, to appear in ADVANCES IN CHEMICAL PHYSICS, and "Conformational Phase Transitions in a Macromolecule: Exactly Solvable Models", by F.W. Wiegel, to appear in Volume 8 of PHASE TRANSITIONS AND CRITICAL PHENOMENA, edited by C. Domb and M.S. Green.

I am indebted to George Bell, Walter Goad and Byron Goldstein of the Los Alamos Laboratory and to Alan Perelson of Brown University for their constructive criticism of the biophysical relevance of the models studied here. I also want to acknowledge discussions with Bob Rubin and Ed DiMarzio of the National Bureau of Standards. I also owe a lot to Piet Mijnlieff for interesting me in the porous medium problem and for tutoring me in some of the chemical physics that is part of it. Last but not least I am indebted to Elly Reimerink for careful typing of the manuscript.

1. INTRODUCTION

The problem of describing the flow of a viscous fluid through a macromolecular system plays a role in several branches of chemical physics and biophysics. As a special case of the more general problem of fluid flow through permeable media its history goes back to Darcy (1856).

For purposes of illustration, consider a system which consists of an incompressible fluid with mass density ρ_0 and viscosity η_0, in a porous medium. The medium might consists of the repeating units of a macromolecule (considered in part B of this monograph) or of a set of cross-linked macromolecules (considered in part C). It will always be assumed that the porous medium moves as a rigid object but this assumption can easily be relaxed by including the equations of motion of the medium in the set of differential equations.

The "microscopic" velocity $\vec{v}(\vec{r},t)$ and pressure $p(\vec{r},t)$ form that solution of the Navier-Stokes equation

$$\rho_0 \frac{\partial \vec{v}}{\partial t} + \rho_0(\vec{v}\cdot\vec{\nabla})\vec{v} = -\vec{\nabla} p + \eta_0 \Delta \vec{v} \tag{1.1}$$

and the incompressibility equation

$$\text{div}\ \vec{v} = 0 \tag{1.2}$$

which obeys the complicated boundary conditions at the highly irregular interface between fluid and medium. These boundary conditions can be "stick" boundary conditions, which require the velocities of fluid and medium to be the same at the interface, or "slip" boundary conditions, which only impose this condition on the component of the velocity normal to the interface. The choice of stick, slip, or mixed boundary conditions depends on the dimensions of the constituents of the porous medium and on other details of the fluid- medium interaction.

Under most conditions of practical interest the non-linear term $\rho_0(\vec{v}\cdot\vec{\nabla})\vec{v}$ in (1.1) can be neglected with respect to the viscosity term $\eta_0\Delta\vec{v}$. The ratio of these terms is of the order of the Reynolds number $\ell v \rho_0/\eta_0$, where ℓ denotes a typical linear dimension of a repeating unit of a macromolecule and v a typical velocity. For a macromolecular coil in water $\ell \cong 10^{-8}$ cm; $\rho_0 \cong 1\,\text{g cm}^{-3}$; $\eta_0 \cong 0.01\,\text{g cm}^{-1}\,\text{s}^{-1}$. With a typical velocity $v \cong 1\,\text{cm s}^{-1}$ the Reynolds number is found to be as small as 10^{-6}. Consequently (1.1) can almost always be replaced by the linearized equation

$$\rho_0 \frac{\partial \vec{v}}{\partial t} = -\vec{\nabla} p + \eta_0 \Delta \vec{v}. \tag{1.3}$$

The unusual properties of life at low Reynolds numbers have been discussed in an elegant paper by Purcell (1977). For related estimates read also Berg and Purcell (1977).

The microscopic fields are rapidly varying functions of the position in space

($\vec{r}$) and time (t) and are, therefore, not accessible to direct observation. Hence $\vec{v}$ and p will not be useful variables to describe the observed motion of the fluid. The "macroscopic" velocity $\vec{V}(\vec{r},t)$ and pressure $P(\vec{r},t)$ are defined as the averages of the corresponding microscopic variables over a small space-time region (ΩT) around ($\vec{r}$,t)

$$\vec{V}(\vec{r},t) \equiv \Omega^{-1}\ T^{-1} \int_\Omega \int_T \vec{v}(\vec{r}\,',t')d^3\vec{r}\,'dt', \qquad (1.4)$$

$$P(\vec{r},t) \equiv \Omega^{-1}\ T^{-1} \int_\Omega \int_T p(\vec{r}\,',t')d^3\vec{r}\,'dt'. \qquad (1.5)$$

Now let the linear dimension of Ω be large as compared to the size (ℓ) of a repeating unit, but small as compared to a characteristic macroscopic length (L)

$$\ell << \Omega^{\frac{1}{3}} << L. \qquad (1.6)$$

If, in the same way, we choose T to be large as compared to the relaxation times associated with a single repeating unit but small as compared to a characteristic time of the imposed macroscopic motion, then the macroscopic fields $\vec{V}$ and P will be slowly varying functions of space and time. It will be shown in the following pages that these macroscopic fields form the solution of a partial differential equation usually connected with the names of Debye, Brinkman and Bueche, who where the first to write it down. In Part A (sections 1 - 7) we shall discuss this fundamental equation and derive it first from irreversible thermodynamics, then from statistical mechanics. The properties of the medium enter into the fundamental equation only through a single phenomenological parameter (the hydrodynamic permeability). The experimental determination of the permeability of a macromolecular system will be discussed in some detail.

Part B (sections 8 - 18) is devoted to the flow of a viscous fluid through and around an isolated macromolecular coil. Here one can essentially distinguish the following three cases. (1) The coil is in a state of uniform rotation with respect to the fluid at infinity. (2) The coil is in a state of uniform translation. (3) The fluid is asymptotically in a state of uniform shear flow. These three cases will lead to the rotational diffusion coefficient, the translational diffusion coefficient and the intrinsic viscosity. The use of the Debye-Brinkman-Bueche equation leads to a considerable simplification of the standard theory of polymer solutions, as expounded, for example, in the book by Yamakawa (1971). A comparison with experiments shows satisfactory agreement in those cases for which data are available.

In part C (sections 19 - 28) the applications of the Debye-Brinkman-Bueche equation to the lateral diffusion of complexes consisting of cross-linked macromolecules in the cell membrane are discussed. First the hydrodynamical properties of cell membranes are reviewed in a qualitative way. The consequences of the Bretscher (1976) flow hypothesis are quantitatively deduced. This is followed by a short study of brownian motion in the presence of a flow field. The discussion of the biophysics of patch- and cap formation demonstrates that Bretscher's hypothesis leads to a quantitative criterion for cap formation. Next, the rotational- and translational lateral diffusion coefficients of a permeable patch of cross-linked immunoglobulins in the cell membrane are calculated. Some space is devoted to a detailed review of the experimental situation.

The macromolecules which play a role in the following sections are long chain-like molecules consisting of a large number of repeating units (monomers). The number of monomers in a single macromolecule is of the order of 10 to 10^5 or larger. The molecular weight of a monomer is of the order 10 to 100; hence the molecular weight of the polymer is of the order 10^2 to 10^7 or larger.

Sometimes only one monomer (A) is present and the polymer has the structure ...-A-A-A-A-... Sometimes the molecule consists of different monomers which can be distributed either periodically or in some random sequence (see, for example, Lehninger 1972 for specific biochemical details).

The linear dimension ℓ of a monomer is of the order of a few Å. Thus, when a polymer consisting of N monomers is fully stretched it can reach a length of the order $N\ell$, i.e. 10^4Å if $N = 10^4$ and $\ell = 1$ Å $= 10^{-8}$cm. This length is of the order of the linear dimension of a small biological cell. Often polymers occur in the shape of a random coil, the linear dimension of which is of the order $N^{\frac{1}{2}}\ell$. For our example this gives 100 Å, which is comparable to the thickness of the membrane around the cell.

Macromolecular systems can be classified roughly into three categories, depending on their concentration. One should of course distinguish between the monomer concentration C_M, defined as the number of monomers per unit volume, and the polymer concentration C_P, defined as the number of polymers per unit volume. If all the polymers in the system consist of the same number (N) of monomers these two concentrations are connected by

$$C_M = NC_P. \tag{1.7}$$

The three concentration regimes are: (a) dilute systems; (b) semidilute systems; (c) concentrated systems.

(a) Dilute systems. In the limit $C_P \to 0$ one has an ideal gas of polymer coils; for small concentrations C_P one has a dilute fluid of coils. Since a polymer coil can be represented roughly by a "cloud" of radius $N^{\frac{1}{2}}\ell$ the monomers concentration inside the coil is of order $C_M^* = N(N^{\frac{1}{2}}\ell)^{-d}$, where $d = 3$ denotes the dimension of space. When the various coils in the system just begin to touch the concentration reaches a value which is comparable to the monomer concentration C_M^* inside a single coil.

Hence the dilute regime is characterized by

$$C_M \ll C_M^* = N^{1-\frac{1}{2}d}\, \ell^{-d}. \tag{1.8}$$

Note that C_M^* is very small for $d = 3$ provided N is large as compared to unity; for $d = 2$, C_M^* does not depend on N.

(b) Semidilute systems. These systems are characterized by overlap between different coils, but the concentration should still be much smaller than the close-packing concentration

$$C_M^* \lesssim C_M \ll \ell^{-d}. \tag{1.9}$$

(c) Concentrated systems. When C_M becomes comparable to ℓ^{-d} the polymers are packed very closely. These system are often locally in a crystalline phase.

2. OPERATIONAL DEFINITION OF THE PERMEABILITY

Operationally the permeability is defined by the following experiment. A cylindrical tube is filled with a porous medium which is kept at rest and which has a constant density. The fluid is pushed through the medium under the influence of a constant pressure gradient $-\vec{\nabla} P$. A stationary state results in which the fluid has a velocity $\vec{V}$ which is practically constant throughout the tube, apart from some small boundary effects (the exact velocity profile will be discussed in section 5). Empirically $\vec{V}$ turns out to be proportional to $-\vec{\nabla} P$ as long as the pressure gradient is not too large. The proportionality factor is written in the form η_0/k, where k is called the hydrodynamic permeability. This leads to the empirical law of Darcy (1856):

$$-\vec{\nabla} P = \frac{\eta_0}{k} \vec{V}. \tag{2.1}$$

The permeability is a function of temperature and of the constitution of fluid and medium. This law should be added to the list of phenomenological equations which, together with the Onsager reciprocal relations, forms the basis of irreversible thermodynamics.

The dimension of the permeability is $[k] = [\text{length}]^2$. Table I gives the permeability of homogeneously distributed poly-α-methylstyrene with mass density $\rho_1 = 1.64 \times 10^{-2}$ g cm^{-3} in cyclohexane and toluene (data taken from Mijnlieff and Jaspers (1971)). It can be seen from these data that for a dilute polymer-solvent mixture, k is roughly of the order $(100Å)^2$; the characteristic length of 100Å which is found here is of the same order of magnitude as the radius of an isolated polymer coil or as the thickness of a biological membrane.

Darcy's law has the following important consequence. If we consider a unit volume of the fluid it will be subject to two forces: (1) a force $-\vec{\nabla} P$ due to the macroscopic pressure in the fluid; (2) a frictional force $\vec{F}$ exerted by the porous medium on the fluid. As the fluid element is not accelerated these two forces must cancel. Hence

$$\vec{F} = +\vec{\nabla} P. \tag{2.2}$$

Combination with Darcy's law gives

$$\vec{F} = -\frac{\eta_0}{k} \vec{V} \tag{2.3}$$

for the force exerted on the fluid by the medium, per unit volume. It should be pointed out that this expression has been derived under the assumption that the porous medium occupies only a negligible fraction of space; the empirical law (2.1) of Darcy holds for any value of this space fraction.

If the porous medium is not at rest, but has a constant macroscopic velocity $\vec{U}$, then the frictional force equals

$$\vec{F} = -\frac{\eta_0}{k}(\vec{V}-\vec{U}). \qquad (2.4)$$

This is easily demonstrated by considering the situation in a frame of reference which moves with a constant velocity $\vec{U}$.

Temp (°C)	$10^{12}\,k(\mathrm{cm}^2)$	Temp (°C)	$10^{12}\,k(\mathrm{cm}^2)$
In cyclohexane		In toluene	
35	1.06	25	0.33
40	0.89	45	0.34
50	0.70	65	0.36
80	0.60	85	0.36
95	0.56	105	0.37
110	0.60	120	0.40
125	0.61		

Table I. Permeability of homogeneous poly-α-methylstyrene at $\rho_1 = 1.64 \times 10^{-2}$ g cm^{-3} in cyclohexane and toluene at different temperatures. From Mijnlieff and Jaspers (1971).

3. RELATION BETWEEN THE PERMEABILITY AND THE SEDIMENTATION COEFFICIENT

From a practical point of view the experiment described in the preceding section cannot be expected to lead to accurate results. In this section we shall, therefore, derive a relation between the permeability and the sedimentation coefficient which leads to more accurate results. This relation was first found by Mijnlieff and Jaspers (1971). The derivation presented here follows Wiegel (1977) and is based on the following fictitious experiment.

Consider a vertical infinite cylindrical tube in which are contained: (1) fluid with mass ρ_0 per unit volume; (2) porous material which is homogeneous on a macroscopic scale and which has mass ρ_1 per unit volume. A constant gravitational field of acceleration g acts along the axis of the cylinder and a constant pressure gradient $-\vec{\nabla} P$ drives the fluid through the tube. Let $\vec{V}$ and $\vec{U}$ denote the average velocities of fluid c.q. medium in the stationary state. We consider this situation in two different frames of reference.

In a frame of reference which moves along the axis of the cylinder with velocity $\vec{U}$ one observes Darcy's experiment. Consequently the frictional force which the fluid exerts on the porous medium per unit volume equals

$$-\vec{F} = \frac{\eta_0}{k}(\vec{V} - \vec{U}) \tag{3.1}$$

according to (2.4).

Next, consider the situation in a frame of reference which moves along the cylinder with a velocity $\vec{W}$ equal to the "mean volume velocity". The mean volume velocity is defined as:

$$\vec{W} = \rho_0 v_0 \vec{V} + \rho_1 v_1 \vec{U}, \tag{3.2}$$

where v_0 denotes the volume of a unit mass of the fluid and v_1 the volume of a unit mass of the medium. Hence $\rho_0 v_0$ denotes the fluid volume per unit volume of the mixture and $\rho_1 v_1$ the volume occupied by the medium per unit volume of the mixture. Obviously

$$\rho_0 v_0 + \rho_1 v_1 = 1. \tag{3.3}$$

In this frame of reference the mean volume velocity vanishes, the fluid flows up with a velocity $\vec{V} - \vec{W}$ and the porous material sediments with a velocity $\vec{U} - \vec{W}$. Hence in this frame of reference one observes a sedimentation experiment in which the sedimentation coefficient s is defined by

$$s \equiv |\vec{U} - \vec{W}|/g. \tag{3.4}$$

The velocity $\vec{U} - \vec{W}$ with which the medium sediments is determined by the balance of the external force and the frictional force (3.1). The external force exerted on the medium in a unit volume equals the force of gravity minus the weight of an equal volume of fluid: $\rho_1 g - \rho_1 v_1 g/v_0$. This gives the equality

$$\rho_1 g(1 - v_1/v_0) = \frac{\eta_0}{k} |\vec{V} - \vec{U}|. \tag{3.5}$$

Elimination of redundant quantities between the last four equations leads to the relation

$$\frac{\eta_0}{k} = \frac{\rho_1}{s} (1 - v_1/v_0)\rho_0 v_0, \tag{3.6}$$

which enables one to calculate the permeability from the sedimentation coefficient. In practical applications $\rho_1 v_1$ will often be very small as compared to unity (at most about 1%); hence $\rho_0 v_0$ is very close to 1 and this factor can be omitted from (3.6). In this monograph it is always assumed that $\rho_1 v_1 \ll 1$, which means that the medium occupies only a negligeable fraction of space.

4. EXPERIMENTAL DETERMINATION OF THE PERMEABILITY

As a consequence of (3.6) a measurement of the permeability amounts to a measurement of the sedimentation coefficient. In the case of homogeneous macromolecular material one has to determine the sedimentation coefficient of homogeneously distributed macromolecular material at some fixed mass density ρ_1.

More specifically, suppose the polymer is poly-α-methylstyrene and the fluid is cyclohexane or toluene. Prepare a sequence of pure samples of this polymer of increasing molecular masses $m_1 < m_2 < m_3 < \dots$ and measure the sedimentation coefficient $s(m_i,\rho_1)$ of the monodisperse polymer of molecular mass m_i at fixed mass density ρ_1.

The quantity

$$s(\rho_1) = \lim_{i\to\infty} s(m_i,\rho_1) \tag{4.1}$$

gives the desired sedimentation coefficient for a uniform distribution of polymeric material at density ρ_1. This is the case because in the limit $m_i \to \infty$ the different polymer coils will overlap and their separate contributions to the density will be washed out, leading to a uniform total density.

The curves $s(m_i,\rho_1)$ were measured by Mijnlieff and Jaspers (1971); their results are qualitatively indicated in figure 1. The resulting curves $k(\rho_1)$ for poly-α-methylstyrene in cyclohexane and toluene are drawn in figure 2. These curves were calculated from the $s(m_i,\rho_1)$ curves using (4.1) and (3.6). The temperature dependence of k has also been measured by these authors and can be found in the publication cited and in Mijnlieff, Jaspers, Ooms and Beckers (1970). Their results are summarized in table I.

A glance at figure 2 shows that the permeability of one and the same polymer is about three times higher in the "poor" solvent cyclohexane than in the "good" solvent toluene. This remarkable solvent effect is due to a local clustering of the repeating units of the polymer when in contact with a poor solvent. As a result the pores between these clusters, through which the solvent has to find its way, become wider, which leads to a higher permeability.

The experimental technique described in this section enables one to measure the value of k fairly accurately. It is instructive to discuss briefly the theoretical expression for k. The theory of Felderhof and Deutch (1975), which we shall review in section 7, predicts the value

$$\frac{\eta_0}{k(\vec{r})} = f\,\nu(\vec{r}), \tag{4.2}$$

where the medium is represented by a collection of mass points, which space density $\nu(\vec{r})$ and a translational friction coefficient f each. If each mass point is represented by a very small sphere of radius a Stokes' formula gives

$$f = 6\pi\eta_0 a. \tag{4.3}$$

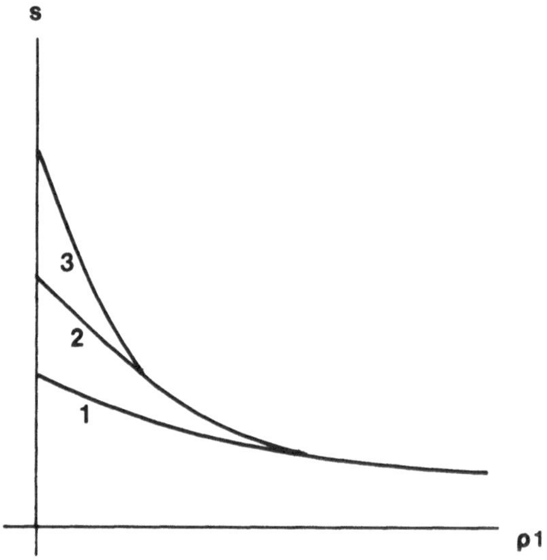

Fig. 1

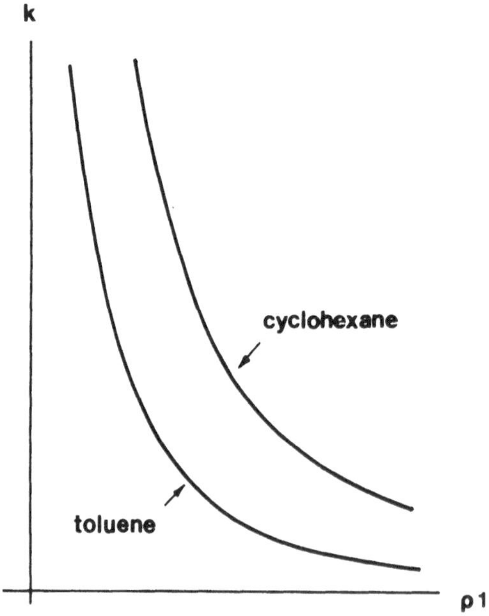

Fig. 2

Experimental results on poly-α-methylstyrene (discussion in section 4)

Combination of the last two equations gives the expression

$$k(\vec{r}) = \{ 6\pi a \nu (\vec{r}) \}^{-1}. \tag{4.4}$$

This formula shows that local clustering of the monomers in a poor solvent can have a considerable effect on the value of the hydrodynamic permeability: as the mass of a cluster of radius a increases proportional to a^3 the number density of clusters $\nu(\vec{r})$ must decrease as a^{-3} in order to keep the total mass density of polymer constant at the value ρ_1. Equation (4.4) indicates that the permeability will increase proportional to a^2.

5. MACROSCOPIC DERIVATION OF THE FUNDAMENTAL EQUATION

Following the results of the last three sections it is now straightforward to derive an equation of motion for the macroscopic velocity $\vec{V}$ and the macroscopic pressure P (compare section 1 for their definition). For these macroscopic fields the Navier-Stokes equation holds in its usual form

$$\rho_0 \frac{\partial \vec{V}}{\partial t} + \rho_0 (\vec{V} \cdot \vec{\nabla}) \vec{V} = - \vec{\nabla} P + \eta_0 \Delta \vec{V} + \vec{F}, \tag{5.1}$$

where $\vec{F}$ denotes the external force exerted on the fluid per unit volume. According to (2.4) the external force is a frictional force which equals

$$\vec{F} = - \frac{\eta_0}{k} (\vec{V} - \vec{U}). \tag{5.2}$$

Substitution of this result gives the fundamental equation

$$\rho_0 \frac{\partial \vec{V}}{\partial t} + \rho_0 (\vec{V} \cdot \vec{\nabla}) \vec{V} = - \vec{\nabla} P + \eta_0 \Delta \vec{V} - \frac{\eta_0}{k} (\vec{V} - \vec{U}), \tag{5.3}$$

which has to be combined with the incompressibility condition:

$$\text{div } \vec{V} = 0. \tag{5.4}$$

The derivation given here is taken from Wiegel and Mijnlieff (1976). Usually one considers flow at low Reynolds numbers (compare our estimate of Reynolds numbers in section 1), in which case the convective term $\rho_0(\vec{V} \cdot \vec{\nabla})\vec{V}$ is negligible as compared to the term $\eta_0 \Delta \vec{V}$. For the stationary state this gives

$$- \vec{\nabla} P + \eta_0 \Delta \vec{V} - \frac{\eta_0}{k} (\vec{V} - \vec{U}) = 0. \tag{5.5}$$

In this form the equation is usually connected with the names of Debye, Brinkman and Bueche.

It was published for the first time by Brinkman (1947a) in a letter to the editor of Physica. This letter was followed by three papers - Brinkman (1947b, 1949a,b) - devoted to important applications. In the following year Debye and Bueche (1948) published an equation similar to (5.5) and results for the intrinsic viscosity and sedimentation coefficient of permeable polymers in solution. They refer to an earlier note by Debye (1947), but this note does not contain any equations at all. Hence, on the basis of his 1947 letter, equation (5.5) should strictly speaking be called the Brinkman equation. It is also remarkable that Debye and Bueche never published the actual derivation of their expressions for the intrinsic viscosity and sedimentation coefficient; this was done for the first time by Felderhof (1975a,b).

In the remaining part of this section we apply the fundamental equation to two simple geometries. In both cases the medium is homogeneous ($k(\vec{r})$ equals a constant k_0)

and at rest ($\vec{U}=0$). In the first example the medium is present in a half-space, the fluid permeates this medium and has a constant velocity V_0 in the rest of space. If the distance between a point of the porous medium and the (macroscopic!) fluid-medium interface is denoted by z the fundamental equation reduces to the form

$$\eta_0 \frac{d^2V}{dz^2} = \frac{\eta_0}{k_0} V, \tag{5.6}$$

with the solution

$$V(z) = V_0 \exp(-z/\sqrt{k_0}). \tag{5.7}$$

This situation would apply roughly to the flow of water through sand underneath the bed of a river. The result shows that the quantity $\sqrt{k_0}$, which has the dimension of a length, measures the distance by which the flow effectively penetrates the medium. We shall, therefore, sometimes call $\sqrt{k_0}$ the penetration length. As shown in section 2, this length is of the order 100Å for typical polymer-solvent mixtures.

The second application of the fundamental equation consists of the quantitative description of the plug flow which we encountered in section 2. In this cylindrical geometry the pressure gradient is a constant. Transforming (5.5) to cylindrical coordinates (r,ϕ,z) around the axis of the tube the cylindrically symmetric solution can be solved from the ordinary differential equation

$$\frac{d^2V}{dr^2} + \frac{1}{r}\frac{dV}{dr} - \frac{V}{k_0} = \frac{1}{\eta_0}\frac{dP}{dz} = \text{constant}, \tag{5.8}$$

where $V(r)$ denotes the z-component of the velocity. The incompressibility equation (5.4) is satisfied automatically. A special solution of the inhomogeneous equation is

$$V = -\frac{k_0}{\eta_0}\frac{dP}{dz} \, . \tag{5.9}$$

The homogeneous equation

$$\frac{d^2V}{dr^2} + \frac{1}{r}\frac{dV}{dr} - \frac{V}{k_0} = 0 \tag{5.10}$$

has the general solution

$$V(r) = A\, I_0\left(\frac{r}{\sqrt{k_0}}\right) + B\, K_0\left(\frac{r}{\sqrt{k_0}}\right), \tag{5.11}$$

where the I_ν and the K_ν denote the modified Bessel functions of order ν. The constant $B=0$ because $V(r)$ has to be finite for $r=0$ and $K_0(\frac{r}{\sqrt{k_0}})$ diverges there. Thus, the general solution of (5.8) reads

$$V(r) = -\frac{k_0}{\eta_0}\frac{dP}{dz} + A\, I_0\left(\frac{r}{\sqrt{k_0}}\right). \tag{5.12}$$

The value of the constant A is determined by the requirement that V should vanish on the wall of the tube ($r=R$). This gives:

$$V(r) = -\frac{k_0}{\eta_0}\frac{dP}{dz}\left\{1-\frac{I_0(r/\sqrt{k_0})}{I_0(R/\sqrt{k_0})}\right\}. \qquad (5.13)$$

For $\sqrt{k_0}$ of the order of 100Å and a macroscopic value of R the resulting velocity will be practically equal to the homogeneous value

$$-\frac{k_0}{\eta_0}\frac{dP}{dz} \qquad (5.14)$$

throughout most of the tube, apart from in a narrow layer close to the wall with a width of the order of the penetration depth $\sqrt{k_0}$. On the other hand, in the free draining regime in which $\sqrt{k_0} \gg R$ we can use eq. 9.6.12 of Abramowitz and Stegun (1970) to expand the modified Bessel function. This leads to

$$V(r) \cong -\frac{1}{4\eta_0}\frac{dP}{dz}(R^2 - r^2), \qquad (R \ll \sqrt{k_0}), \qquad (5.15)$$

which is Poiseulle's formula for the flow of a viscous fluid through a cylindrical tube.

6. THE OSEEN TENSOR

In the next section a microscopic derivation of the fundamental equation is presented. As a preliminary to that presentation this section is devoted to the following problem: Consider an incompressible Newtonian fluid at low Reynolds numbers, under the influence of an external force field $\vec{F}(\vec{r})$ of the form

$$\vec{F}(\vec{r}) = +\,\delta(\vec{r}-\vec{r}_0)\vec{1}, \tag{6.1}$$

where $\vec{1}$ denotes some unit vector and where $\vec{r}_0$ denotes the point where the force "works". We want to calculate the resulting velocity and pressure fields in the stationary state.

In order to solve the time independent linearized Navier-Stokes equation

$$-\vec{\nabla} p + \eta_0 \Delta \vec{v} + \delta(\vec{r}-\vec{r}_0)\vec{1} = 0 \tag{6.2}$$

one proceeds as if the pressure were a given function of $\vec{r}$. In this case it is convenient first to solve the simpler scalar equation

$$\Delta G = \delta(\vec{r}). \tag{6.3}$$

Writing

$$G(\vec{r}) = (2\pi)^{-3} \int \tilde{G}(\vec{k}) \exp\,(i\,\vec{k}\cdot\vec{r})\; d^3\vec{k}, \tag{6.4}$$

$$\delta(\vec{r}) = (2\pi)^{-3} \int \exp\,(i\,\vec{k}\cdot\vec{r})\; d^3\vec{k}, \tag{6.5}$$

one finds upon substitution into (6.3)

$$\tilde{G}(\vec{k}) = -\,|\vec{k}|^{-2}. \tag{6.6}$$

Hence $G(\vec{r})$ is given by the Fourier transform

$$G(\vec{r}) = -(2\pi)^{-3} \int |\vec{k}|^{-2} \exp(+i\,\vec{k}\cdot\vec{r})\; d^3\vec{k} = \frac{-1}{4\pi r}\,, \tag{6.7}$$

where we used eq. 4.3.142 of Abramowitz and Stegun (1970) to evaluate the integral. As a consequence of this result the solution of the equation

$$\Delta\phi = f(\vec{r}), \tag{6.8}$$

where $f(\vec{r})$ denotes an arbitrary function which vanishes sufficiently rapidly at infinity, is given by

$$\phi(\vec{r}) = -\frac{1}{4\pi} \int |\vec{r}-\vec{r}\,'|^{-1} f(\vec{r}\,')\; d^3\vec{r}\,'. \tag{6.9}$$

Now consider the original problem (6.2). As all three components can be written in the form (6.8) the solution is

$$v_j(\vec{r}) = + \frac{I_j}{4\pi\eta_0 |\vec{r} - \vec{r}_0|} - \frac{1}{4\pi\eta_0} \int |\vec{r} - \vec{r}\,'|^{-1} \frac{\partial p(\vec{r}\,')}{\partial r'_j} d^3\vec{r}\,', \qquad (6.10)$$

where $j = 1,2,3$ denotes the component along the Cartesian axis. The pressure, which still appears as an unknown quantity on the right hand side of this equation, can be determined from the condition of incompressibility

$$0 = \operatorname{div} \vec{v} = + \frac{1}{4\pi\eta_0} \operatorname{div}(|\vec{r} - \vec{r}_0|^{-1} \vec{I})$$
$$- \frac{1}{4\pi\eta_0} \int p(\vec{r}\,') \, \Delta' \, |\vec{r} - \vec{r}\,'|^{-1} d^3\vec{r}\,', \qquad (6.11)$$

where a partial integration was performed and where Δ' denotes the Laplace operator with respect to $\vec{r}\,'$. With the use of (6.3) and (6.7) this equation can be written in the form

$$\frac{1}{\eta_0} p(\vec{r}) = \frac{-1}{4\pi\eta_0} \operatorname{div} (|\vec{r} - \vec{r}_0|^{-1} \vec{I}). \qquad (6.12)$$

Consequently the pressure is given by the expression

$$p(\vec{r}) = + \sum_j Q_j(\vec{r} - \vec{r}_0) \, I_j, \qquad (6.13)$$

where the Oseen vector Q has components

$$Q_j(\vec{R}) = \frac{R_j}{4\pi |\vec{R}|^3}, \qquad (j = 1,2,3). \qquad (6.14)$$

The velocity field is found upon substitution of the pressure into (6.10). Performing two partial integrations one obtains

$$v_j(\vec{r}) = \frac{I_j}{4\pi\eta_0 |\vec{r} - \vec{r}_0|} + \frac{1}{16\pi^2\eta_0} \frac{\partial}{\partial r_j} \frac{\partial}{\partial \vec{r}} \cdot \vec{I} \int |\vec{r} - \vec{r}\,'|^{-1} \, |\vec{r}\,' - \vec{r}_0|^{-1} d^3\vec{r}\,'. \qquad (6.15)$$

In order to evaluate the integral imagine a plane through $\vec{r}$ and $\vec{r}_0$ and introduce polar coordinates $\rho \equiv |\vec{r}\,' - \vec{r}_0|, \phi$ in this plane which are centered at $\vec{r}_0$. With a cut-off at some very large but constant value ρ_0 of the polar radius vector one finds

$$\int |\vec{r} - \vec{r}\,'|^{-1} \, |\vec{r}\,' - \vec{r}_0|^{-1} d^3\vec{r}\,' =$$
$$= 2\pi \int_0^{\rho_0} \rho d\rho \int_0^{\pi} (\rho^2 + |\vec{r} - \vec{r}_0|^2 - 2 \, |\vec{r} - \vec{r}_0| \, \rho \cos\phi)^{-\frac{1}{2}} \sin\phi \, d\phi$$
$$= 4\pi\rho_0 - 2\pi |\vec{r} - \vec{r}_0|. \qquad (6.16)$$

When this result is substituted back into (6.15) the dependence on ρ_0 drops out as a result of the differentiations. Hence one can formally take the limit $\rho_0 \to \infty$ and find

$$v_j(\vec{r}) = \frac{I_j}{4\pi\eta_0|\vec{r}-\vec{r}_0|} - \frac{1}{8\pi\eta_0}\left\{\frac{I_j}{|\vec{r}-\vec{r}_0|} - \sum_{k=1}^{3} I_k \frac{(r_k-r_{0,k})(r_j-r_{0,j})}{|\vec{r}-\vec{r}_0|^3}\right\}. \qquad (6.17)$$

This is usually written in the form

$$v_j(\vec{r}) = \sum_k T_{j,k}(\vec{r}-\vec{r}_0)\, I_k, \qquad (6.18)$$

where the Oseen tensor $\overleftrightarrow{T}$ has components

$$T_{j,k}(\vec{R}) = \frac{1}{8\pi\eta_0}\left\{\delta_{j,k} + \frac{R_jR_k}{|\vec{R}|^3}\right\}. \qquad (6.19)$$

Note that the pressure field drops off proportional to R^{-2} at large distances, but that the velocity field is proportional only to R^{-1}.

The Oseen vector $\vec{Q}$ and the Oseen tensor $\overleftrightarrow{T}$ describe the hydrodynamic interactions between mass points in a viscous fluid. For more complicated models the hydrodynamic interaction has been analyzed also. The interested reader is referred especially to three papers by Jones (1978a,b,c) and a paper by Reuland, Felderhof and Jones (1978) in which the hydrodynamic interaction between two permeable spherically symmetric polymers is studied.

7. MICROSCOPIC DERIVATION OF THE FUNDAMENTAL EQUATION

We are now in a position to derive the fundamental equation (5.5) from the microscopic equations of motion of the field. This was done for the first time by Felderhof and Deutch (1975), whose method will be followed here.

The porous medium is represented by a collection of mass points located at positions $\vec{r}_j$ and with frictional coefficients f_j. Note that this model assumes that only a negligible fraction of space is occupied by the porous medium. The mass points move with a prescribed velocity $\vec{U}(\vec{r})$. Let $\vec{v}_0(\vec{r})$ and $p_0(\vec{r})$ denote the velocity and pressure in the absence of the mass points; for small Reynolds numbers these fields are the solution of

$$-\vec{\nabla} p_0 + \eta_0 \Delta \vec{v}_0 = 0, \tag{7.1}$$

$$\text{div } \vec{v}_0 = 0, \tag{7.2}$$

which obeys the proper boundary conditions. In the presence of the mass points a total force with space density

$$\vec{f}(\vec{r}) = \sum_j \vec{F}_j \, \delta(\vec{r} - \vec{r}_j), \tag{7.3}$$

$$\vec{F}_j = f_j \, [\vec{U}(\vec{r}_j) - \vec{v}(\vec{r}_j)] \tag{7.4}$$

works on the fluid. The velocity $\vec{v}(\vec{r})$ equals the sum of the unperturbed field $\vec{v}_0(\vec{r})$ and the perturbations due to all the mass points. Using (6.18) this gives

$$\vec{v}(\vec{r}_j) = \vec{v}_0(\vec{r}_j) + \sum_k \overleftrightarrow{T}(\vec{r}_j - \vec{r}_k) f_k \, [\vec{U}(\vec{r}_k) - \vec{v}(\vec{r}_k)]. \tag{7.5}$$

Combination of the last two equations gives

$$\vec{F}_j = f_j \, [\vec{U}(\vec{r}_j) - \vec{v}_0(\vec{r}_j)] - \sum_k f_j \, \overleftrightarrow{T}(\vec{r}_j - \vec{r}_k) \vec{F}_k. \tag{7.6}$$

This equation has to be combined with the microscopic equation of motion

$$-\vec{\nabla} p + \eta_0 \Delta \vec{v} + \vec{f}(\vec{r}) = 0, \tag{7.7}$$

$$\text{div } \vec{v} = 0. \tag{7.8}$$

In order to obtain the macroscopic description of the system one has to study the behavior of the average quantities

$$\vec{V}(\vec{r}) = \langle \vec{v}(\vec{r}) \rangle, \tag{7.9}$$

$$P(\vec{r}) = \langle p(\vec{r}) \rangle , \tag{7.10}$$

$$\vec{F}(\vec{r}) = \langle \vec{f}(\vec{r}) \rangle , \tag{7.11}$$

where the average $\langle \rangle$ is taken over all those different ways to distribute the point masses which are consistent with the macroscopic distribution of mass. Taking the average of (7.7) and (7.8) gives

$$-\vec{\nabla} P + \eta_0 \Delta \vec{V} + \vec{F}(\vec{r}) = 0, \tag{7.12}$$

$$\operatorname{div} \vec{V} = 0. \tag{7.13}$$

This set of equations still has to be "closed" by espressing $\vec{F}(\vec{r})$ in terms of $\vec{V}(\vec{r})$.

Multiplying (7.6) with $\delta(\vec{r} - \vec{r}_j)$ and summing over j gives the exact relation

$$\vec{f}(\vec{r}) = [\vec{U}(\vec{r}) - \vec{v}_0(\vec{r})] \sum_j f_j \, \delta(\vec{r} - \vec{r}_j) - \sum_j f_j \, \delta(\vec{r} - \vec{r}_j) \sum_k \vec{\vec{T}}(\vec{r} - \vec{r}_k) \vec{F}_k . \tag{7.14}$$

Now the last term can be written in the continuous notation

$$\sum_k \vec{\vec{T}}(\vec{r} - \vec{r}_k) \vec{F}_k = \int \vec{\vec{T}}(\vec{r} - \vec{r}\,') \vec{f}(r') d^3\vec{r}\,' . \tag{7.15}$$

When (7.14) is averaged one finds

$$\vec{F}(\vec{r}) = [\vec{U}(\vec{r}) - \vec{v}_0(\vec{r})] \langle \sum_j f_j \, \delta(\vec{r} - \vec{r}_j) \rangle - \langle \sum_j f_j \, \delta(\vec{r} - \vec{r}_j) \int \vec{\vec{T}}(\vec{r} - \vec{r}\,') f(\vec{r}\,') d^3\vec{r}\,' \rangle . \tag{7.16}$$

The first term on the right side leads to the microscopic definition of the permeability

$$\frac{\eta_0}{k(\vec{r})} \equiv \langle \sum_j f_j \, \delta(\vec{r} - \vec{r}_j) \rangle . \tag{7.17}$$

In the special case in which the mass points are small spheres with radius a and space density $\nu(\vec{r})$ the last equation predicts a permeability

$$k(\vec{r}) = \{ 6 \pi a \, \nu(\vec{r}) \}^{-1} ; \tag{7.18}$$

compare our comments at the end of section 4.

The average of the product in the second term of (7.16) is approximated by the product of the averages

$$< \sum_j f_j \, \delta(\vec{r}-\vec{r}_j) \int \vec{\vec{T}}(\vec{r}-\vec{r}\,')\vec{f}(\vec{r}\,')d^3\vec{r}\,'>$$

$$\cong < \sum_j f_j \, \delta(\vec{r}-\vec{r}_j)> \int \vec{\vec{T}}(\vec{r}-\vec{r}\,')< \vec{f}(\vec{r}\,')> d^3\vec{r}\,'. \tag{7.19}$$

This ad-hoc approximation is analogous to the mean-field approximation in equilibrium statistical mechanics. A further simplification of this theory has been studied by Deutch and Felderhof (1975).

Combination of the last four equations gives the integral equation

$$\vec{F}(\vec{r}) = \frac{\eta_0}{k(\vec{r})} [\vec{U}(\vec{r})-\vec{v}_0(\vec{r})] - \frac{\eta_0}{k(\vec{r})} \int \vec{\vec{T}}(\vec{r}-\vec{r}\,')\vec{F}(\vec{r}\,')d^3\vec{r}\,'. \tag{7.20}$$

At this point in the derivation it should be noted that equation (7.12) has already been formally solved in the previous section. The solution is found by adding to $\vec{v}_0(\vec{r})$ an integral of an expression of the form (6.18); in a continuous notation

$$\vec{V}(\vec{r}) = \vec{v}_0(\vec{r}) + \int \vec{\vec{T}}(\vec{r}-\vec{r}\,')\vec{F}(\vec{r}\,')d^3\vec{r}\,'. \tag{7.21}$$

Combination of the last two equations gives

$$\vec{F}(\vec{r}) = \frac{\eta_0}{k(\vec{r})} [\vec{U}(\vec{r}) - \vec{V}(\vec{r})] \tag{7.22}$$

and substitution of this result back into (7.12) gives the fundamental equation in the form

$$-\vec{\nabla} P + \eta_0 \Delta \vec{V} + \frac{\eta_0}{k(\vec{r})} [\vec{U}(\vec{r}) - \vec{V}(\vec{r})] = 0. \tag{7.23}$$

This equation is identical to (5.5) which resulted from irreversible thermodynamics. In the same way the expression (7.22) for the force exerted on the fluid is identical to (2.,4). Before we turn to some important applications of the equation we want to remind the reader that three assumptions have been made in the derivation of this section: (1) the relevant Reynolds numbers are small as compared to unity; (2) the fraction of space occupied by the medium is small as compared to unity; (3) the mean-field approximation (7.19) is valid.

8. THE EINSTEIN RELATIONS

In the next six sections the diffusion coefficients of an isolated macromolecule in a viscous fluid will be calculated. The Einstein relation

$$D_T = \frac{k_B T}{f_T} \qquad (8.1)$$

relates the translational diffusion coefficient D_T to the translational friction coefficient f_T, which is defined as the drag force on the macromolecule per unit relative velocity. Boltzmann's constant is denoted by k_B and the absolute temperature by T. In the same way the Einstein relation

$$D_R = \frac{k_B T}{f_R} \qquad (8.2)$$

relates the rotational diffusion coefficient D_R to the rotational friction coefficient f_R, defined as the torque on the coil per unit angular velocity. Because of these relations the determination of the diffusion coefficients amounts to the determination of the corresponding friction coefficients.

Relation (8.1) was first derived by Einstein (1905, 1956). A slightly modernized version of his derivation runs as follows. Consider a large number of these polymers, without interactions, in a cylindrical volume, under the influence of an external force K which acts along the axis of the cylinder. In thermodynamic equilibrium the spacial density $\nu(x)$ of the polymers will be a function of the coordinate x along the axis of the cylinder, which can be found in the following way. As a result of the force K the particles will have an average velocity K/f_T along the axis of the cylinder; this results in a flux $K\nu(x)/f_T$. Also, because of the definition of the diffusion coefficient, the density gradient $d\nu/dx$ leads to a second flux $-D_T d\nu/dx$. Hence the density obeys the equation

$$\frac{K\nu}{f_T} - D_T \frac{d\nu}{dx} = 0, \qquad (8.3)$$

which has the solution:

$$\nu(x) = \nu(0) \exp\left(\frac{Kx}{f_T D_T}\right). \qquad (8.4)$$

But statistical mechanics gives the equilibrium distribution in the form

$$\nu(x) = \nu(0) \exp\left(\frac{Kx}{k_B T}\right). \qquad (8.5)$$

Identification of the last two equations gives the Einstein relation (8.1). The second relation can be proved in a similar way.

From a theoretical point of view the Einstein relations are very handy because with their help the calculation of a diffusion coefficient (a statistical quantity) is immediately reduced to the calculation of the friction coefficient which is a hydrodynamic quantity.

Experimentally the translational diffusion coefficient is determined by measuring the average squared displacement $\langle \vec{r}^2 \rangle$ of the polymer during a time interval t and applying the relation

$$\langle \vec{r}^2 \rangle = 2\, d\, D_T\, t, \tag{8.6}$$

where d denotes the dimension of space. For a typical polymer in a solvent at room temperature D_T is of the order 10^{-7} cm^2 s^{-1} and f_T is of the order 5×10^{-7} g s^{-1}. The polymer will diffuse over a distance of 100Å (the order of its radius of gyration) in a time of the order 10^{-6} s. The interested reader is referred to a paper by Berg and Purcell (1977) for order of magnitude estimations of diffusion processes which play a role in the theory of chemoreception.

In the following sections we calculate f_T and f_R for an isolated macromolecule by representing the macromolecular coil by a rigid but porous sphere with a permeability which is a function k(r) of the radial distance to the center of the sphere. The validity of this so called porous sphere model has recently been discussed extensively by Mijnlieff and Wiegel (1978).

9. ROTATIONAL DIFFUSION COEFFICENT: GENERAL THEORY

Imagine a porous macromolecular coil which rotates with constant angular velocity ω_0 around the z-axis of a Cartesian set of coordinates. The velocity components of the sphere are

$$U_1 = -\omega_0 y, \tag{9.1}$$

$$U_2 = +\omega_0 x, \tag{9.2}$$

$$U_3 = 0, \tag{9.3}$$

where $\vec{U} \equiv (U_1,U_2,U_3)$. In the stationary state the velocity $\vec{V} \equiv (V_1,V_2V_3)$ of the fluid is the solution of (5.4) and (5.5). One expects that the pressure will be constant throughout the volume and that each fluid elements rotates around the z-axis. This type of flow automatically satisfies the incompressibility condition (5.4). The x and y components of the velocity have to be solved from

$$\Delta V_i - \frac{1}{k} V_i = -\frac{1}{k} U_i \qquad (i = 1,2). \tag{9.4}$$

It will turn out to be convenient to introduce spherical coordinates (r,θ,ϕ) by

$$x = r \sin\theta \cos\phi, \tag{9.5}$$

$$y = r \sin\theta \sin\phi, \tag{9.6}$$

$$z = r \cos\theta. \tag{9.7}$$

In these variables the Laplace operator becomes

$$\Delta = \frac{\partial^2}{\partial r^2} + \frac{2}{r}\frac{\partial}{\partial r} + \frac{1}{r^2}\frac{\partial^2}{\partial \theta^2} + \frac{\mathrm{cotg}\theta}{r^2}\frac{\partial}{\partial \theta} + \frac{1}{r^2 \sin^2\theta}\frac{\partial^2}{\partial \phi^2} . \tag{9.8}$$

Denoting the angular velocity of the fluid at a distance r from the origin by $\omega(r)$ the velocity components are

$$V_1(r,\theta) = -r\,\omega(r) \sin\theta \sin\phi , \tag{9.9}$$

$$V_2(r,\theta) = +r\,\omega(r) \sin\theta \cos\phi. \tag{9.10}$$

Substitution into (9.4) shows that these equations give the solution provided the unknown function $\omega(r)$ is the solution of the ordinary differential equation

$$\left[\frac{d^2}{dr^2} + \frac{2}{r}\frac{d}{dr} - \frac{2}{r^2} - \frac{1}{k(r)} \right] (r\omega) = -\frac{\omega_0 r}{k(r)} . \tag{9.11}$$

The boundary conditions are $\omega(\infty) = 0$ and $|\omega(0)| < \infty$.

In the next section this equation will be solved for a coil of uniform permeability. In this section we show that the rotational diffusion coefficient is determined uniquely by the asymptotic behavior at large distances of the solution of (9.11).

Outside the coil the permeability is infinite, so the equation simplifies to

$$\left[\frac{d^2}{dr^2} + \frac{2}{r}\frac{d}{dr} - \frac{2}{r^2}\right] (r\omega) = 0. \tag{9.12}$$

This equation has a solution of the form $\omega \sim r^{\alpha}$ with $\alpha = -3$ or 0. Because of the boundary condition at infinity the second solution has to be rejected and one has the asymptotic solution

$$\omega(r) = \frac{A\omega_0}{r^3} \qquad \text{(outside coil)}, \tag{9.13}$$

which holds outside the coil. The constant A has to be determined from the behavior of the solution inside the coil.

The total torque which has to be exerted on the coil to keep it in a state of uniform rotation is equal to

$$\vec{\tau} = \int \vec{r} \times \vec{F} \ d^3\vec{r}, \tag{9.14}$$

where $\vec{F}$ is given by (5.2). The only non-vanishing component is τ_3. Substituting (5.2) and (5.5) one finds

$$\tau_3 = \eta_0 \int (y \Delta V_1 - x \Delta V_2) \ d^3\vec{r}. \tag{9.15}$$

Using Green's theorem the volume integral can be written as a surface integral

$$\tau_3 = \eta_0 \oint (y \frac{\partial V_1}{\partial n} - x \frac{\partial V_2}{\partial n} - V_1 \frac{\partial y}{\partial n} + V_2 \frac{\partial x}{\partial n}) \ d^2S. \tag{9.16}$$

This integral can be extended over any surface which completely contains the coil. For this surface we choose a sphere with some large radius r. Using (9.9), (9.10) and (9.13) one finds

$$\tau_3 = 3\eta_0\omega_0 \ A \ r^{-2} \oint \sin^2\theta \ d^2S. \tag{9.17}$$

As $d^2S = r^2 \sin\theta \ d\theta \ d\phi$ the surface integral can be performed in a straightforward way. The result is

$$\tau_3 = 8\pi\eta_0\omega_0 A. \tag{9.18}$$

Hence the rotational friction coefficient is

$$f_R = \frac{\tau_3}{\omega_0} = 8\pi\eta_0 A \qquad (9.19)$$

and the rotational diffusion coefficient equals:

$$D_R = \frac{k_B T}{8\pi\eta_0 A}, \qquad (9.20)$$

where the second Einstein relation (8.2) was used. The elegant result (9.19) was published for the first time by Felderhof and Deutch (1975). It shows that the rotational diffusion coefficient is only determined by the asymptotic behavior of the flow pattern at large distances from the coil.

For the record, we note here the dimensions of these transport coefficients: $[f_R]$ = [energy] [time] and $[D_R] = [\text{time}]^{-1}$.

An interesting general result has also been derived by Felderhof and Jones (1978) who prove a Faxén theorem for the force and torque exerted by the fluid on a permeable spherically symmetric macromolecule, using the most general slip-stick boundary conditions. Related work has been reported by Jones (1978b). Felderhof (1976a,b) has studied the concentration dependence of f_R for a suspension of permeable macromolecules.

10. ROTATIONAL DIFFUSION COEFFICIENT OF A UNIFORM SPHERE

The simplest and most important application of the general theory of the preceeding section is to the case where the macromolecular coil is represented by a sphere of radius R and constant permeability k_0

$$k(r) = k_0 \qquad (r < R), \tag{10.1}$$

$$= \infty \qquad (r > R). \tag{10.2}$$

The finite discontinuity in $k(r)$ at $r = R$ will generally lead to finite discontinuities in the second derivatives of the velocity and in the first derivatives of the pressure (compare eq. (5.5)). The pressure, velocity and the first derivatives of the velocity should be continuous at $r = R$.

For $r > R$ the solution has already been given in (9.13)

$$\omega(r) = \frac{\omega_0 A}{r^3} \qquad (r > R). \tag{10.3}$$

For $r < R$ one has to solve

$$\left[\frac{d^2}{dr^2} + \frac{2}{r}\frac{d}{dr} - \frac{2}{r^2} - \frac{1}{k_0}\right](r\omega) = -\frac{\omega_0}{k_0} r. \tag{10.4}$$

A special solution of the inhomogeneous equation is given by

$$\omega(r) = \omega_0. \tag{10.5}$$

The substitutions

$$r = \xi\sqrt{k_0}, \tag{10.6}$$

$$r\omega = f(\xi), \tag{10.7}$$

transform the homogeneous equation into the equation

$$\frac{d^2 f}{d\xi^2} + \frac{2}{\xi}\frac{df}{d\xi} - \left(1 + \frac{2}{\xi^2}\right) f = 0. \tag{10.8}$$

The solution which vanishes for $\xi \downarrow 0$ is

$$f(\xi) = B\sqrt{k_0}\left(\frac{\cosh\xi}{\xi} - \frac{\sinh\xi}{\xi^2}\right), \tag{10.9}$$

where B is a constant. Hence inside the coil the solution is given by

$$\omega(r) = \omega_0 + B\left(\frac{k_0}{r^2}\cosh\frac{r}{\sqrt{k_0}} - \frac{k_0\sqrt{k_0}}{r^3}\sinh\frac{r}{\sqrt{k_0}}\right). \tag{10.10}$$

At $r = R$ both ω and ω' must be continuous. This leads to two conditions

$$\frac{\omega_0 A}{R^3} = \omega_0 + B\left(\frac{\cosh\sigma}{\sigma^2} - \frac{\sinh\sigma}{\sigma^3}\right), \tag{10.11}$$

$$-\frac{3\omega_0 A}{R^4} = \frac{B}{\sqrt{k_0}}\left(\frac{\sinh\sigma}{\sigma^2} - 3\,\frac{\cosh\sigma}{\sigma^3} + 3\,\frac{\sinh\sigma}{\sigma^4}\right), \tag{10.12}$$

where

$$\sigma = \frac{R}{\sqrt{k_0}} \tag{10.13}$$

is a dimensionless quantity which measures the ratio between the radius R and the distance $\sqrt{k_0}$ by which the fluid effectively penetrates the porous medium (compare the discussion in section 5). Solving (10.11) and (10.12) one finds for the two constants the explicit values

$$A = R^3\left(1 + \frac{3}{\sigma^2} - \frac{3}{\sigma}\,\mathrm{cotgh}\,\sigma\right), \tag{10.14}$$

$$B = -\frac{3\omega_0\sigma}{\sinh\sigma}\;. \tag{10.15}$$

Using (9.19) and (9.20) one finds the rotational friction coefficient

$$f_R = 8\pi\eta_0 R^3\left(1 + \frac{3}{\sigma^2} - \frac{3}{\sigma}\,\mathrm{cotgh}\,\sigma\right), \tag{10.16}$$

and the rotational diffusion coefficient

$$D_R = \frac{k_B T}{8\pi\eta_0 R^3}\left(1 + \frac{3}{\sigma^2} - \frac{3}{\sigma}\,\mathrm{cotgh}\,\sigma\right)^{-1}. \tag{10.17}$$

These results were first found by Felderhof and Deutch (1975).

It is of some interest to discuss the two limiting cases $\sigma\to\infty$ and $\sigma\to 0$ of these formulae. In the limit $\sigma\to\infty$ the porous sphere becomes an impermeable sphere and one obtains the rotational friction coefficient of a hard sphere with stick boundary conditions

$$f_R(\infty) = 8\pi\eta_0 R^3. \tag{10.18}$$

If $\sigma \ll 1$ the expression (10.16) simplifies to

$$f_R \simeq \frac{8}{15}\pi\eta_0 R^3\sigma^2 \qquad (\sigma \ll 1), \tag{10.19}$$

where we used eq. 4.5.67 of Abramowitz and Stegun (1970). The correctness of this formula can be verified independently using the following argument. For $\sigma \ll 1$ little interaction exists between the fluid and the porous sphere, hence $\vec{V}$ will be negligible as compared to $\vec{U}$ and (5.2) gives

$$\vec{F} \cong \frac{\eta_0}{k} \vec{U}. \qquad (10.20)$$

Substituting this approximation into (9.14) one finds for the torque:

$$\tau_3 \cong \frac{\eta_0}{k_0} \int (xU_2 - yU_1) d^3\vec{r}, \qquad (10.21)$$

where the integration is restricted to the interior of the sphere. When (9.1,2) are substituted into the right-hand side of this equation the torque is found to equal

$$\tau_3 \cong \omega_0 \frac{\eta_0}{k_0} \frac{8}{15} \pi R^5; \qquad (10.22)$$

this leads to the result (10.19). The regime $\sigma \ll 1$ is called the free draining regime for reasons which will be obvious from this derivation.

In table II the correction factor in (10.16) due to the finite permeability of the sphere has been calculated using table 4.15 of Abramowitz and Stegun (1970). Experimentally σ is typically of order unity and neither the free-draining limit nor the impermeable sphere limit can be used.

Felderhof and Deutch (1975) also calculate the rotational friction coefficient of a hollow spherical shell with a constant permeability at the surface. A biophysical application of these results can be found in a paper by McCammon, Deutch and Felderhof (1975). Experimental results for D_R are scarce. This probably accounts for the fact that the general formalism of section 9 has not been applied to more realistic models for the macromolecular coil, like the Gaussian model to be discussed shortly.

σ	$G_1(\sigma)$
0	0
1	0.0609
2	0.1940
3	0.3284
4	0.4370
5	0.5195
6	0.5833
7	0.6327
8	0.6719
9	0.7037
10	0.7300
∞	1

Table II. Values of the correction factor $G_1(\sigma) = (1 + \frac{3}{\sigma^2} - \frac{3}{\sigma} \text{cotgh}\,\sigma)$ in eq. (10.16) due to a finite permeability of the porous sphere. Note the slow convergence to the limit value 1 for large values of σ.

11. TRANSLATIONAL DIFFUSION COEFFICIENT: GENERAL THEORY

The calculation of the translational diffusion coefficient is similar to the calculation in section 9, but the Ansatz to be made is somewhat less straightforward. When an impermeable sphere of radius R moves through a viscous fluid with constant relative velocity v_0 the drag force on the sphere is given by Stokes' formula

$$F = 6\pi\eta_0 R v_0. \tag{11.1}$$

This gives a translational friction coefficient

$$f_T = 6\pi\eta_0 R, \tag{11.2}$$

and a translational diffusion coefficient

$$D_T = \frac{k_B T}{6\pi\eta_0 R}. \tag{11.3}$$

These formulae are often used to assign an "effective hydrodynamic radius" to a macromolecule the diffusion coefficient of which has been measured. However, a polymer in solution cannot be represented by an impermeable sphere because the solvent can flow through as well as around the coil. Hence the flow has to be solved from the fundamental equation (5.5) in conjunction with (5.4). In this section we shall present the details of this calculation for the general case of a porous coil, following Felderhof (1975a) and Wiegel and Mijnlieff (1977b).

Consider an isolated macromolecule at rest in the origin of a Cartesian system of coordinates (x,y,z). In the absence of the coil the fluid would be in a state of uniform flow with a velocity

$$\vec{v} = (0,0,v_0) \tag{11.4}$$

and the pressure would equal a constant p_0 throughout the fluid. Owing to the presence of the coil the actual velocity and pressure will be the solutions of (5.4) and (5.5), with $\vec{U} = 0$, which approach the unperturbed fields at large distances from the origin. For these solutions one makes the Ansatz

$$P = p_0 - \eta_0 \frac{\xi(r)}{r} (\vec{r}\cdot\vec{v}), \tag{11.5}$$

$$\vec{V} = \psi(r)\,\vec{v} - \nu(r)\,\vec{r}\times(\vec{r}\times\vec{v}), \tag{11.6}$$

where $\xi(r)$, $\psi(r)$ and $\nu(r)$ are unknown scalar functions of the radial distance to the origin. This Ansatz, which was first used by Felderhof (1975a), is inspired by the following observations: (a) the pressure is a scalar and the velocity a vector; (b) the velocity and the excess pressure should be linear in $\vec{v}$; (c) the only vectors in the problem are $\vec{r}$ and $\vec{v}$. These conditions lead uniquely to (11.5) and somewhat limit the possibilities for $\vec{V}$.

It is tedious but straightforward to verify by substitution that the Ansatz solves (5.4) and (5.5) provided the three unknown functions are the solutions of the three coupled ordinary differential equations

$$\nu = \frac{1}{2r}\psi', \tag{11.7}$$

$$\psi'' + \frac{4}{r}\psi' - k^{-1}\psi + \xi' = 0, \tag{11.8}$$

$$\xi'' + \frac{2}{r}\xi' - \frac{2}{r^2}\xi - (k^{-1})'\psi = 0, \tag{11.9}$$

where the prime denotes differentiation with respect to r. The boundary conditions are that velocity and pressure stay finite in the origin and approach the unperturbed fields if $r \to \infty$. This implies

$$\psi(0) \quad \text{finite}, \tag{11.10}$$

$$\xi(0) \quad \text{finite}, \tag{11.11}$$

$$\psi(\infty) = 1, \tag{11.12}$$

$$\xi(\infty) = 0. \tag{11.13}$$

As in the case of the rotational diffusion coefficient one shows easily that the translational diffusion coefficient is determined by the asymptotic behavior of the solution at large distances. Outside the coil $k^{-1} = 0$ and (11.8,9) become

$$\psi'' + \frac{4}{r}\psi' + \xi' = 0, \tag{11.14}$$

$$\xi'' + \frac{2}{r}\xi' - \frac{2}{r^2}\xi = 0. \tag{11.15}$$

The last equation, with boundary condition (11.13), has the solution

$$\xi(r) = \frac{C}{r^2} \qquad \text{(outside coil)}. \tag{11.16}$$

Substituting into (11.14) and using the boundary condition (11.12) one finds

$$\psi(r) = 1 - \frac{C}{r} + \frac{D}{r^3} \qquad \text{(outside coil)}, \tag{11.17}$$

where C and D are constants to be determined shortly.

Now the z-component of the total force which the fluid exerts on the coil follows from (5.2)

$$-F_3 = \eta_0 \int k^{-1}(\vec{r})\ V_3\ (\vec{r})\ d^3\vec{r}. \tag{11.18}$$

The integration is over all space. Using (5.5) this can be written as

$$\begin{aligned} -F_3 &= \int (-\frac{\partial P}{\partial z} + \eta_0 \Delta V_3) d^3\vec{r} \\ &= \oint [-P \frac{z}{r} + \eta_0 (\frac{x}{r} \frac{\partial V_3}{\partial x} + \frac{y}{r} \frac{\partial V_3}{\partial y} + \frac{z}{r} \frac{\partial V_3}{\partial z})]\ d^2S. \end{aligned} \tag{11.19}$$

In the second equality we used Gauss' theorem to replace the volume integral by a surface integral over the surface of a sphere with some large radius r. Substituting (11.16) and (11.17) into (11.5) and (11.6) gives the solution

$$P = p_0 - \eta_0 v_0 C \frac{z}{r^3}, \tag{11.20}$$

$$V_3 = v_0 - \frac{v_0 C}{2r} - \frac{v_0 C z^2}{2r^3} + O(r^{-3}) \tag{11.21}$$

outside the coil. If this is combined with (11.19) and if the limit $r \to \infty$ is taken one finds

$$-F_3 = 4\pi\, \eta_0 C\, v_0 \tag{11.22}$$

for the drag force on the coil. Consequently the translational friction coefficient is given by

$$f_T = 4\pi\, \eta_0 C, \tag{11.23}$$

and the translational diffusion coefficient by

$$D_T = \frac{k_B T}{4\pi\eta_0 C}. \tag{11.24}$$

In the next section this formalism will be applied to the uniform coil. The dimensions of these important transport coefficients are: $[f_T]$ = [mass]. $[\text{time}]^{-1}$ and $[D_T]$ = $[\text{length}]^2\ [\text{time}]^{-1}$.

Further general results on the translational diffusion coefficient are given by Felderhof and Jones (1978) who prove a Faxén theorem for the drag force exerted on a permeable macromolecule. Jones (1978b) has calculated f_T using the hydrodynamic interaction of permeable coils and Felderhof (1976a,b) studied the concentration dependence of f_T for a suspension of permeable polymers.

12. TRANSLATIONAL DIFFUSION COEFFICIENT OF A UNIFORM SPHERE

For the uniform sphere model of section 10 the derivative $(k^{-1})'$ of the inverse permeability vanishes both inside and outside the sphere. This implies that in this case the solution of (11.9) is given by

$$\xi(r) = \frac{C}{r^2} \qquad (r < R), \qquad (12.1)$$

$$= Er \qquad (r < R), \qquad (12.2)$$

where E is a constant. The function $\psi(r)$ is given by (11.17) outside the sphere and has to be solved from

$$\psi'' + \frac{4}{r}\psi' - k_0^{-1}\psi + E = 0 \qquad (12.3)$$

inside the sphere.

The solution of this equation which is finite at r=0 is

$$\psi(r) = Ek_0 + F\left(\frac{1}{r^2}\cosh\frac{r}{\sqrt{k_0}} - \frac{\sqrt{k_0}}{r_0^3}\sinh\frac{r}{\sqrt{k_0}}\right) \qquad (r < R), \qquad (12.4)$$

as can be verified by substitution.

The boundary conditions at r=R have been discussed in section 10. The continuity of the pressure implies, through (11.5), that $\xi(r)$ should be continuous in R. The continuity of the velocity and its first derivatives implies, through (11.6) and (11.7) that ψ, ψ' and ψ'' should be continuous in R. This gives the set of relations

$$\frac{C}{R^2} = ER, \qquad (12.5)$$

$$1 - \frac{C}{R} + \frac{D}{R^3} = Ek_0 + F\left(\frac{1}{R^2}\cosh\sigma - \frac{\sqrt{k_0}}{R^3}\sinh\sigma\right), \qquad (12.6)$$

$$\frac{C}{R^2} - \frac{3D}{R^4} = F\left\{-\frac{3}{R^3}\cosh\sigma + \left(\sigma + \frac{3}{\sigma}\right)\frac{1}{R^3}\,\mathrm{sh}\,\sigma\right\}, \qquad (12.7)$$

$$-\frac{2C}{R^3} + \frac{12D}{R^5} = F\left\{\frac{1}{R^4}(12+\sigma^2)\cosh\sigma - \frac{1}{R^4}\left(5\sigma + \frac{12}{\sigma}\right)\sinh\sigma\right\}, \qquad (12.8)$$

where σ was defined in (10.13). It is straightforward, but tedious, to solve the four constants C, D, E, F from these four equations. One finds

$$C = \frac{3}{2}RG_0(\sigma)\left\{1 + \frac{3}{2\sigma^2}G_0(\sigma)\right\}^{-1}, \qquad (12.9)$$

where

$$G_0(\sigma) = 1 - \frac{1}{\sigma}\,\mathrm{tgh}\,\sigma. \qquad (12.10)$$

The translational friction coefficient now follows from (11.23)

$$f_T = 6\pi\eta_0 R\, G_0(\sigma) \{ 1 + \frac{3}{2\sigma^2} G_0(\sigma)\}^{-1} , \tag{12.11}$$

and the diffusion coefficient is given by the first Einstein relation

$$D_T = \frac{k_B T}{6\pi\eta_0 R G_0(\sigma)} \{ 1 + \frac{3}{2\sigma^2} G_0(\sigma)\} . \tag{12.12}$$

These results were first given by Debye and Bueche (1948) but the first derivation was published by Felderhof (1975a), who also treated the spherical shell. In the impermeable sphere limit $\sigma \to \infty$ the function $G_0(\sigma)$ tends to unity and one recovers Stokes' formulae (11.2) and (11.3). In the free draining regime one finds $G_0(\sigma) \cong \frac{1}{3}\sigma^2$ and

$$f_T \cong \frac{4}{3}\pi\eta_0\, R\, \sigma^2, \qquad (\sigma \ll 1). \tag{12.13}$$

This result can be verified immediately with the use of lowest order perturbation theory as discussed in section 10. In table III we have tabulated the correction factor $G_0(\sigma)\ \{1 + \frac{3}{2\sigma^2} G_0(\sigma)\}^{-1}$ in (12.11) due to the finite permeabilty of the sphere.

The translational diffusion coefficient has alo been calculated by Jones, Felderhof and Deutch (1975) for a polymer which is constrained to move along the interface between two fluids. These authors consider two polymer models: rigid rods oriented at right angles to the interface and porous spheres with their center in the interface and with a permeability which is constant in each hemisphere.

The uniform porous sphere, although better than the impermeable sphere, is still somewhat unsatisfactory as a model for a real macromolecular coil. In the next two sections we shall discuss a more realistic model (the Gaussian model) and calculate its translational diffusion coefficient.

σ	$G_0(\sigma)\ \{ 1 + \frac{3}{2\sigma^2} G_0(\sigma)\}^{-1}$
0	0
1	0.1756
2	0.4337
3	0.6013
4	0.7009
5	0.7634
6	0.8054
7	0.8352
8	0.8574
9	0.8745
10	0.8880
∞	1

Table III. Values of the correction factor $G_0(\sigma)\ \{ 1 + \frac{3}{2\sigma^2} G_0(\sigma)\}^{-1}$ in eq. (12.11) due to the finite permeability of the sphere.

13. THE FREE RANDOM WALK MODEL AND THE GAUSSIAN COIL

In a real isolated macromolecule in a solvent the permeability will be some function $k(r)$ of the radial distance (r) to the center of mass of the coil. This function is related in a complicated way to the local mass density $\rho_1(r)$ of macromolecular material (compare the discussion in section 4). A fairly accurate form of the curve $k(r)$ can be found by: (1) using free random walk statistics to determine the function $\rho_1(r)$; (2) using the experimentally determined curve $k(\rho_1)$. In this section we shall first derive the main consequences of the free random walk model for macromolecular chain statistics. Next, we shall discuss the Gaussian coil. The section ends with some general considerations concerning the conformational phase transitions in a macromolecular chain and with comments on the role of space dimension in the escape probability and the excluded volume effect.

Whereas the successive repeating units of a macromolecule are strongly bound, so that their mutual distance is fixed, in many cases the units can be easily rotated with respect to one another. These two features of real macromolecules form the basis of the random walk model for macromolecular statistics, in which a configuration of a macromolecule consisting of N repeating units is represented by a random walk consisting of N steps, each of length ℓ, where ℓ equals the average distance between two successive repeating units in the macromolecule. Depending on the interactions between the repeating units of the macromolecule one should impose certain constraints on these random walks. In first approximation such effects are neglected altogether; consequently the random walks considered here are free. In this free random walk model the length of each step is constant, and no correlations exist between the directions of successive steps. At the end of this section we shall consider briefly the case in which strong repulsive interactions between the monomers are taken into account (the excluded volume problem; also compare the Appendix).

Denote the probability density that a free random walk which starts at the origin will reach the point $\vec{r}$ after N steps by $P(\vec{r},N)$. The first of these probability distributions is

$$P(\vec{r},1) = (4\pi\ell^2)^{-1}\,\delta(|\vec{r}| - \ell), \tag{13.1}$$

and the higher ones can be found from the first one by an integration over the N-1 coordinates $\vec{r}_1,\vec{r}_2,\ldots,\vec{r}_{N-1}$ of the endpoints of the 1st, 2nd,...,$(N-1)^{st}$ step

$$P(\vec{r},N) = \int d^3\vec{r}_1 \int d^3\vec{r}_2 \ldots \int d^3\vec{r}_{N-1} \prod_{i=0}^{N-1} (4\pi\ell^2)^{-1}\delta(|\vec{r}_{i+1} - \vec{r}_i| - \ell). \tag{13.2}$$

Here $\vec{r}_0 = \vec{0}$, $\vec{r}_N = \vec{r}$ and $\delta(x)$ denotes Dirac's delta function.

The Fourier transform of $P(\vec{r},1)$, which will be denoted by $\tilde{P}(\vec{k},1)$, can be calculated directly

$$\tilde{P}(\vec{k},1) = \int (4\pi\ell^2)^{-1}\ \delta(|\vec{r}|-\ell)e^{i\vec{k}.\vec{r}}\ d^3\vec{r} = \frac{\sin k\ell}{k\ell}\ , \tag{13.3}$$

where k denotes the length of the vector $\vec{k}$. As $P(\vec{r},N)$ is related to $P(\vec{r},1)$ by an N-1 fold convolution product its Fourier transform is the Nth power of $\tilde{P}(\vec{k},1)$

$$\tilde{P}(\vec{k},N) = \int P(\vec{r},N)e^{i\vec{k}.\vec{r}}\ d^3\vec{r} = \left(\frac{\sin k\ell}{k\ell}\right)^N. \quad (N = 1,2,3...) \tag{13.4}$$

Of course the Fourier transform contains the same information as the probability distribution itself, because an inverse Fourier transform

$$\tilde{P}(\vec{r},N) = (2\pi)^{-3} \int \left(\frac{\sin k\ell}{k\ell}\right)^N e^{-i\vec{k}.\vec{r}}\ d^3\vec{k} \quad (N = 1,2,3...) \tag{13.5}$$

gives an integral representation for the probability distribution.

Instead of the explicit representation (13.5) for the probability distribution of free random walks, it is often convenient to use the asymptotic form of this exact expression for $N \gg 1$. This can be obtained from the asymptotic form of the Fourier transforms for $k\ell \ll 1$. Expanding (13.3) for small $k\ell$ one finds

$$\begin{aligned}\tilde{P}(\vec{k},1) &= 1 - \frac{1}{6}k^2\ell^2 + O(k^4\ell^4)\\ &= e^{-\frac{1}{6}k^2\ell^2} + O(k^4\ell^4).\end{aligned} \tag{13.6}$$

Substituting this expansion into (13.4) gives the asymptotic form of the Fourier transform

$$\tilde{P}(\vec{k},N) \simeq e^{-\frac{1}{6}Nk^2\ell^2}\ . \qquad (N\gg1) \tag{13.7}$$

Taking the inverse Fourier transform and performing the integration over $\vec{k}$ one finds the asymptotic form of the probability distribution

$$P(\vec{r},N) \simeq \left(\tfrac{2}{3}\pi N\ell^2\right)^{-3/2} e^{-3\vec{r}^2/2N\ell^2}. \qquad (N\gg1) \tag{13.8}$$

The last equation displays the well known Gaussian character of the probability distribution of free random walks, a direct consequence of the central limit theorem of probability theory. For a further discussion of random walks the reader should consult Chandrasekhar (1943), Wax (1954) or Barber and Ninham (1970).

It is straightforward to derive the corresponding results for random walks in a plane. For future reference we need only quote the asymptotic forms of the probability distribution and its Fourier transform. Instead of (13.6) one now has

$$\tilde{P}(\vec{k},1) = 1 - \tfrac{1}{4}k^2\ell^2 + O(k^4\ell^4)$$
$$= e^{-\frac{1}{4}k^2\ell^2} + O(k^4\ell^4). \tag{13.9}$$

This gives the asymptotic form of the Fourier transform of the probability distribution

$$\tilde{P}(\vec{k},N) = e^{-\frac{1}{4}Nk^2\ell^2}, \qquad (N>>1) \tag{13.10}$$

and the asymptotic form of the probability distribution itself

$$P(\vec{r},N) \cong (\pi N\ell^2)^{-1} e^{-\vec{r}^2/N\ell^2} . \qquad (N>>1) \tag{13.11}$$

Whereas the last three results for walks in the plane are quite similar to the corresponding results for random walks in space, there is a difference between the case of dimension 2 and the case of dimension 3 that has significance to the type of conformational phase transitions which occur in macromolecules. This will be discussed at the end of this section.

Note that both in two and in three dimensions the quantity

$$\rho_g = \ell\sqrt{N} \tag{13.12}$$

is the root of the average squared end to end distance. The reader is warned that this quantity is often called the "radius of gyration" in the chemical physical literature, contrary to the formal definition of the radius of gyration in classical mechanics.

Returning now to the free random walk model eq. (13.8) it turns out that, even for this simple case, the distribution of mass around the center of gravity has not been calculated analytically (compare the discussion in section 8 of Yamakawa (1971)). A fair approximation is

$$\rho_1(\vec{r}) = mP(\vec{r},N), \tag{13.13}$$

where m denotes the total mass of the macromolecule. Combining this with a $k(\rho_1)$ curve which qualitatively has the form indicated in figure 2 one finds that the k(r) curve can be approximated by the expression

$$k(r) = K \exp(Qr^2), \tag{13.14}$$

where K and Q are positive constants. An isolated macromolecular coil with this distribution of the permeability will be called a Gaussian coil. The determination of the constants K and Q, which are different for every polymer-solvent pair, has been discussed by Mijnlieff and Wiegel (1978). An appropriate choice is to identify

K with the permeability

$$K \cong K\,(\rho_1(0)) \tag{13.15}$$

which corresponds with the density

$$\rho_1(0) = m\,(\tfrac{2}{3}\,\pi\rho_g^2)^{-3/2} \tag{13.16}$$

in the center of the coil, and to take

$$Q \cong \frac{3}{2\rho_g^2}\,. \tag{13.17}$$

The values of m and ρ_g can be determined empirically for every polymer-solvent pair from sedimentation and light scattering experiments The first application of the Gaussian coil model to the theory of the transport properties of a dilute polymer solution is the paper by Ooms, Mijnlieff and Beckers (1970) in which for a few specific polymer-solvent pairs D_T is calculated with a purely numerical procedure.

A warning is in place here. Under certain conditions a macromolecule can go through a configurational phase transition of some kind of another. This would lead to replacement of the random coil (13.8) by a helix, a folded conformation, a globule or some other drastically different conformation. These conformational phase transitions have recently been discussed in detail by Wiegel (1979d); they all lead to the breakdown of both the free random walk model and the Gaussian coil.

We end this section with a discussion of the role of the dimension of space. The peculiar effect of space dimension (d) is most easily demonstrated in a calculation of the escape probability. Suppose we imagine an infinitesimal volume element $d\Omega$ around the origin and ask for the probability U_N that a random walk which starts from the origin has returned to some point inside $d\Omega$ after N steps. This probability is given according to (13.8) and (13.11) by

$$U_N \cong \begin{cases} (\pi N \ell^2)^{-1}\, d\Omega, & (d=2) \\ (\tfrac{2}{3}\,\pi N \ell^2)^{-3/2}\, d\Omega. & (d=3) \end{cases} \tag{13.18}$$

U_N gives the probability that the endpoint $\vec{r}_N$ of the last step of the random walk belongs to $d\Omega$, regardless of whether or not any of the endpoints $\vec{r}_1, \vec{r}_2, \ldots, \vec{r}_{N-1}$ of the intermediate steps belonged to $d\Omega$. Let V_N denote the probability that $\vec{r}_N$ belongs to $d\Omega$ but that none of the endpoints of intermediate steps belonged to $d\Omega$. Thus V_N gives the probability that the random walk enters $d\Omega$ for the first time at the N th step; whereas U_N gives the probability that the random walk is in $d\Omega$ after N steps, maybe after several previous visits. We want to calculate V_N.

The relation between U_N and V_N can be found if one orders the random walks which are in $d\Omega$ after N steps according to the number of the $\vec{r}_1, \vec{r}_2, \ldots, \vec{r}_{N-1}$ which belonged to $d\Omega$

$$U_N = V_N + \sum_{N_1+N_2=N}{}' V_{N_1} V_{N_2} + \sum_{N_1+N_2+N_3=N}{}'' V_{N_1} V_{N_2} V_{N_3} + \ldots . \tag{13.19}$$

For example the second term on the right hand side gives the contribution to U_N of all random walks which have visited $d\Omega$ twice: once after N_1 steps, then again after another N_2 steps. The constraints on the $N_1, N_2, \ldots$ which are indicated under the summation signs make it very difficult to solve the unknown V_N from the known U_N. The method of generating functions has especially been designed to eliminate constraints of the type which occur in the last equation. Introducing the generating functions

$$U(z) \equiv \sum_{N=1}^{\infty} U_N z^N , \tag{13.20}$$

$$V(z) \equiv \sum_{N=1}^{\infty} V_N z^N , \tag{13.21}$$

multiplying both sides of (13.19) with z^N and summing N from 1 to ∞, one finds

$$U(z) = V(z) + V^2(z) + V^3(z) + \ldots = \frac{V(z)}{1 - V(z)} . \tag{13.22}$$

From this relation V(z) can be solved in terms of U(z)

$$V(z) = \frac{U(z)}{1 + U(z)} , \tag{13.23}$$

and V_N can be calculated by complex integration

$$V_N = (2\pi i)^{-1} \oint_C \frac{U(z)}{1 + U(z)} z^{-N-1} dz \tag{13.24}$$

where the contour C encircles the origin of the complex z-plane once in counter-clockwise direction.

In order to demonstrate the role of the dimension of the space in which the random walk proceeds we use the relations (13.20 - 24) to calculate the probability W_0 that a random walk of infinite length will ever return to the volume element $d\Omega$ in which it originated. Obviously, W_0 is the sum of the probabilities V_N to return to $d\Omega$ for the first time after N steps. According to (13.21) this sum can be expressed in the generating function V(z) for $z = 1$

$$W_0 = \sum_{N=1}^{\infty} V_N = V(1). \tag{13.25}$$

Using (13.23) one finds

$$W_0 = \frac{U(1)}{1+U(1)} \; . \tag{13.26}$$

Now with the explicit expressions (13.11) and (13.8) one has

$$U(1) = (\pi\ell^2)^{-1}\, d\Omega \sum_{N=1}^{\infty} N^{-1} = \infty, \qquad (d=2) \tag{13.27}$$

$$U(1) = (\tfrac{2}{3}\pi\ell^2)^{-3/2}\, d\Omega \sum_{N=1}^{\infty} N^{-3/2} < \infty . \qquad (d=3) \tag{13.28}$$

Substituting these results into (13.26) one finds

$$W_0 = 1, \qquad (d=2) \tag{13.29}$$

$$W_0 < 1. \qquad (d=3) \tag{13.30}$$

Consequently, a continuing random walk in a plane will always return to a narrow vicinity $d\Omega$ of the point where it started, but a random walk in three-dimensional space has a finite probability $(1-W_0) > 0$ to "escape" from $d\Omega$ forever. This influence of dimension on the escape probability can qualitatively be understood as follows. A macromolecular coil consisting of N repeating units, each of length ℓ, can roughly be represented by a cloud with a radius of the order $N^{\frac{1}{2}}\ell$. Hence the probability to return to the vicinity $d\Omega$ of $\vec{r}_0$ after N steps drops off with increasing N as $N^{-\frac{1}{2}d}$ and U(1) is proportinal to $\sum_{N=1}^{\infty} N^{-\frac{1}{2}d}$. This series diverges if $d \le 2$ and converges if $d > 2$. Hence, according to (13.26) the escape probability $1-W_0$ will be 0 if $d \le 2$ and > 0 if $d > 2$. Note that these qualitative estimates are also meaningful if d is interpreted as a real continuous parameter.

Finally, we want to point out that excluded volume effects lead to some corrections to the free random walk statistics (13.8), but these corrections - although important from a conceptual point of view - are negligeably small from a quantitative point of view. This problem has recently been reviewed in detail by McKenzie (1976) and by Lifshitz, Grosberg and Khokhlov (1978). Until now the excluded volume problem has resisted all attempts at an analytic solution, although there are some indications that the two-dimensional problem can be solved analytically (Wiegel, 1979e). It can also be shown in the following way that the excluded volume effect is of no consequence for $d > 4$. In the free random walk model the interactions between the different monomers are ignored. Let us estimate qualitatively when this is permitted. Assume that two monomers have a large positive interaction energy E_0 when their distance is smaller than some threshold distance. The monomer density in the ideal coil will be of the order $N^{1-\frac{1}{2}d}\,\ell^{-d}$, hence the energy of self-interaction will be of the order $E_0 N^{2-\frac{1}{2}d}$. For $\bar{N} \gg 1$ this will be very small if $d > 4$ but it will be very large if $d < 4$. Hence, the excluded volume effect, due to the self-interaction, can be neglected if $d > 4$ but has to be taken into account for $d < 4$.

The considerations in this section show that two dimensions play a special role in the statistical mechanics of isolated macromolecules: $d = 2$ above which the escape probability becomes non-zero, and $d = 4$ below which the excluded volume effect has to be taken into account. The real world, with $d = 3$, seems to be the most complex of all possible worlds. Further comments on the excluded volume problem have been collected in the Appendix.

14. TRANSLATIONAL DIFFUSION COEFFICIENT OF A GAUSSIAN COIL

In this section the general formalism of section 11 will be applied to the model for which

$$k(r) = K \exp(Qr^2). \tag{14.1}$$

We shall follow the paper by Wiegel and Mijnlieff (1977b). When (14.1) is substituted into (11.8) and (11.9) one finds equations which can be brought into a dimensionless form by introducing the dimensionless variables

$$x = r\sqrt{Q}, \tag{14.2}$$

$$h(x) = \psi(r), \tag{14.3}$$

$$q(x) = \frac{\xi(r)}{\sqrt{Q}}. \tag{14.4}$$

The dimensionless equations are

$$\frac{d^2h}{dx^2} + \frac{4}{x}\frac{dh}{dx} - \alpha e^{-x^2} h + \frac{dq}{dx} = 0, \tag{14.5}$$

$$\frac{d^2q}{dx^2} + \frac{2}{x}\frac{dq}{dx} - \frac{2}{x^2} q + 2\alpha x e^{-x^2} h = 0, \tag{14.6}$$

where

$$\alpha = K^{-1}Q^{-1}. \tag{14.7}$$

The boundary conditions (11. 12, 13) imply $h(\infty) = 1$, $q(\infty) = 0$. The other boundary conditions (11.10) and (11.11) imply that $q(0)$ and $h(0)$ should be finite. As $h(0)$ is finite the last term on the left hand side of (14.6) vanishes for $x \downarrow 0$. Consequently the behavior of $q(x)$ close to the origin can be solved from

$$\frac{d^2q}{dx^2} + \frac{2}{x}\frac{dq}{dx} - \frac{2}{x^2} q \cong 0 \qquad (x \downarrow 0). \tag{14.8}$$

The solution which gives a finite value for $q(0)$ is x; this gives the alternative boundary condition

$$q(0) = 0 \tag{14.9}$$

and $(dq/dx)_{x=0}$ will be finite. Using this information in (14.5) one finds that $h(x)$ behaves near the origin like the solution of

$$\frac{d^2h}{dx^2} + \frac{4}{x}\frac{dh}{dx} \cong \text{constant} \qquad (x \downarrow 0). \tag{14.10}$$

The solution of this equation which gives a finite value for h(0) is $c_1 + c_2 x^2$, hence

$$\left(\frac{dh}{dx}\right)_{x=0} = 0. \tag{14.11}$$

The boundary conditions (14.9) and (14.11) turn out to be much easier to use that the two original ones.

Whereas an analytic solution of (14.5,6) seems impossible to obtain a numerical solution can be found in the following way. Using (11.16) and (11.17) one has for the solution outside the coil

$$q(x) = \frac{c}{x^2}, \tag{14.12}$$

$$h(x) = 1 - \frac{c}{x} + \frac{d}{x^3}, \tag{14.13}$$

where c and d are constants to be determined. We may use the last two results for $x > x_0$ where x_0 is some large positive constant. Then use (14.5,6) to calculate q(0) and h'(0) numerically. In general the values found will violate (14.9) and/or (14.11). The values of c and d are now adjusted till both boundary conditions are satisfied within some error margin.

Denoting the correct value of c by $\alpha\Psi(\alpha)$ one has

$$C = c\,Q^{-\frac{1}{2}} = Q^{-\frac{1}{2}}\,\alpha\Psi(\alpha). \tag{14.14}$$

Using the general results (11.23,24) one finds

$$f_T = 4\pi\eta_0\,Q^{-\frac{1}{2}}\,\alpha\Psi(\alpha), \tag{14.15}$$

$$D_T = \frac{k_B T Q^{\frac{1}{2}}}{4\pi\eta_0\alpha\Psi(\alpha)} \tag{14.16}$$

The values of the function $\Psi(\alpha)$ have been calculated in this way for α up to 17; the results, which were first published by Wiegel and Mijnlieff (1977a,b), are listed in table IV. The Gaussian model leads to values of D_T which are in satisfactory agreement with experimental data; this will be the subject of section 18.

α	$\Psi(\alpha)$	$\Phi(\alpha)$
0	0.443	0.332
1	0.359	0.321
2	0.303	0.311
3	0.263	0.301
4	0.233	0.292
5	0.209	0.284
6	0.190	0.277
7	0.175	0.270
8	0.162	0.263
9	0.151	0.257
10	0.141	0.251
11	0.133	0.246
12	0.125	0.241
13	0.119	0.236
14	0.113	0.231
15	0.108	0.227
16	0.103	0.223
17	0.099	0.219

Table IV. The functions $\Psi(\alpha)$ and $\Phi(\alpha)$ which determine the translational diffusion coefficient and the intrinsic viscosity of Gaussian coils, according to eqs. (14.16) and (17.11).

15. VISCOSITY: GENERAL THEORY

In a celebrated paper Einstein (1906) calculated the viscosity of a dilute suspension of impermeable spheres. For purposes of comparison, the interested reader is also referred to Einstein (1911, 1956). Einstein's expression is often used to interpret the viscosity of a solution of macromolecules in terms of an "effective hydrodynamic radius". However, the same criticism applies here as was applied in the case of the translational diffusion coefficient. In this section we calculate the intrinsic viscosity of permeable macromolecules in solution, following Felderhof (1975b) and Wiegel and Mijnlieff (1977b).

Let the macromolecule be located at the origin of coordinates. At large distances from the coil the fluid is in a state of uniform shear flow

$$\vec{v} = (G_0 y, 0, 0), \tag{15.1}$$

with shear rate G_0. The coil, which is represented by a rigid porous sphere, will rotate around the z-axis as a result of the interaction with the fluid

$$\vec{U} = (-\omega y, \omega x, 0). \tag{15.2}$$

The angular velocity ω can be found by imposing the requirement that in the stationary state the total torque of the forces which the fluid exerts on the coil should vanish. It can be shown that this implies that the angular velocity equals half the shear rate

$$\omega = -\tfrac{1}{2} G_0. \tag{15.3}$$

For the actual velocity and pressure fields one makes the Ansatz

$$P = p_0 - \eta_0 \frac{\chi(r)}{r^2} (\vec{r} \cdot \vec{v}), \tag{15.4}$$

$$\vec{V} = \vec{U} + \phi(r)(\vec{v}-\vec{U}) - \mu(r)\vec{r} \times (\vec{r} \times [\vec{v}-\vec{U}]), \tag{15.5}$$

where $\chi(r)$, $\phi(r)$ and $\mu(r)$ are unknown scalar functions of the distance to the origin of the coil. This Ansatz, attributable to Felderhof (1975b), is similar to eqs. (11.5,6). After substitution and a tedious, somewhat unelegant calculation, one finds the solution provided

$$\mu = \frac{1}{3r} \phi', \tag{15.6}$$

$$\phi'' + \frac{6}{r} \phi' - k^{-1} \phi + \frac{\chi'}{r} = 0, \tag{15.7}$$

$$\chi'' + \frac{2}{r} \chi' - \frac{6}{r^2} \chi - r(k^{-1})' \phi = 0. \tag{15.8}$$

The boundary conditions are that V and P are finite in the origin and approach the unperturbed fields at large distances; this implies

$$\phi(0) \quad \text{finite}, \tag{15.9}$$

$$\chi(0) = 0, \tag{15.10}$$

$$\phi(\infty) = 1, \tag{15.11}$$

$$\lim_{r\to\infty} \frac{\chi(r)}{r} = 0. \tag{15.12}$$

For $r \to \infty$ the permeability goes to infinity and the fourth term on the left-hand side of (15.8) vanishes. The resulting equation has the solution

$$\chi(r) = \frac{2A}{r^3} \qquad \text{(outside coil)}. \tag{15.13}$$

Upon substitution into eq. (15.7) one finds

$$\phi(r) = 1 - \frac{A}{r^3} + \frac{B}{r^5} \qquad \text{(outside coil)}, \tag{15.14}$$

where the boundary conditions were used and where the third term on the left-hand side of (15.7) was set equal to zero. By substituting into (15.5) one finds the asymptotic form of the three components of the velocity

$$V_1 = - G_0 A \frac{x^2 y}{r^5} + G_0 y + O(r^{-4}), \tag{15.15}$$

$$V_2 = - G_0 A \frac{x y^2}{r^5} + O(r^{-4}), \tag{15.16}$$

$$V_3 = - G_0 A \frac{xyz}{r^5} + O(r^{-4}), \tag{15.17}$$

which formulae determine the viscosity in the following way.

Consider a dilute solution which contains n_p of these macromolecules per unit volume. Let the solution be contained between two parallel walls situated at $y = \pm L$ and let these walls move with equal but opposite velocities ($\pm GL$) along the x-axis in such a way that the macroscopic velocity field has a shear rate G. We calculate the viscosity of the solution following the method of Burgers (1938). The macroscopic velocity field should be distinguished from the local velocity field in the vicinity of a macromolecule. The velocity field which would be found at the position ($\vec{r}_i$) of a particular macromolecule, when that molecule has first been removed, has a shear rate G_0. The presence of the molecule at $\vec{r}_i$ will add to this unperturbed velocity a small correction given by (15.15-17), and these corrections have to be summed over all the coils in the fluid.

To be more specific, we calculate the correction to the velocity in some point (x,y,z) due to all coils present in a thin slice of fluid parallel to the x,z-plane and with thickness dy'. As the coils are distributed with number density n_p the correction to V_1 can be found from the integral

$$\Delta V_1 = -G_0 A n_p dy' \int_{-\infty}^{\infty} dx' \int_{-\infty}^{\infty} dz'(x-x')^2(y-y')|\vec{r}-\vec{r}'|^{-5}$$
$$= -\frac{2}{3}\pi G_0 A n_p \frac{y-y'}{|y-y'|} dy'. \tag{15.18}$$

Note that $\partial\Delta V_1/\partial y = 0$, so the molecules in this slice do not change the local shear rate; therefore, the shear at the upper and lower plates still equals G_0. Hence, the x-component of the force which the fluid exerts on a unit area of the upper surface equals $\eta_0 G_0$. By integrating (15.18) over y' and adding the local velocity field one finds the macroscopic velocity field $\vec{V}^{(M)}$ described by

$$V_1^{(M)} = Gy, \tag{15.19}$$

$$V_2^{(M)} = 0, \tag{15.20}$$

$$V_3^{(M)} = 0. \tag{15.21}$$

Carrying out the integration one finds

$$G = (1 - \frac{4}{3}\pi n_p A)\, G_0. \tag{15.22}$$

The viscosity η of the solution is operationally defined by measuring the force per unit area and dividing by the shear rate. But as the force per unit area has an x-component equal to $\eta_0 G_0$ this gives

$$\eta G = \eta_0 G_0. \tag{15.23}$$

Combining the last two equations one finds for the relative increase of the viscosity

$$\frac{\eta-\eta_0}{\eta_0} = \frac{4}{3}\pi n_p A. \tag{15.24}$$

This relation expresses the viscosity of a dilute solution of permeable macromolecules in terms of the asymptotic behavior of the flow field around one macromolecule.

In this section we have followed the method of Burgers (1938) and Hermans (1953) to calculate the viscosity of the solution from the asymptotic flow fields (15.15-17). Other methods are due to Einstein (1906, 1911, 1956), Kramers (1946), Kirkwood and Riseman (1948), Yamakawa (1971), Landau and Lifshitz (1959) and Peterson and Fixman

(1963). Several of these methods have been discussed by Felderhof (1975b). Felderhof (1976a,b) also has studied the concentration dependence of the increase in the viscosity of a suspension of spherically symmetric polymers.

16. VISCOSITY OF A DILUTE SOLUTION OF UNIFORMLY POROUS SPHERES

For the uniform sphere model, discussed in sections 10 and 12, the derivative $(k^{-1})'$ vanishes inside and outside the sphere. This immediately gives the fields

$$\chi(r) = \frac{2A}{r^3} \qquad (r > R), \tag{16.1}$$

$$\phi(r) = 1 - \frac{A}{r^3} + \frac{B}{r^5} \qquad (r > R). \tag{16.2}$$

Inside the coil the solution of (15.8) is

$$\chi(r) = Dr^2 \qquad (r < R), \tag{16.3}$$

where we used the boundary condition (15.10). The function ϕ inside the coil has to be solved from

$$\phi'' + \frac{6}{r}\phi' - k_0^{-1}\phi + 2D = 0. \tag{16.4}$$

The solution of this equation which is finite at $r = 0$ is given by

$$\phi(r) = 2Dk_0 + C\ \{(\frac{r}{\sqrt{k_0}})^{-3} \sinh(\frac{r}{\sqrt{k_0}}) - 3(\frac{r}{\sqrt{k_0}})^{-4} \cosh(\frac{r}{\sqrt{k_0}}) + 3(\frac{r}{\sqrt{k_0}})^{-5} \sinh(\frac{r}{\sqrt{k_0}})\} \qquad (r < R), \tag{16.5}$$

as can be verified by substitution.

The continuity of χ, ϕ, ϕ' and ϕ'' at $r = R$ leads to four identities

$$\frac{2A}{R^3} = DR^2, \tag{16.6}$$

$$1 - \frac{A}{R^3} + \frac{B}{R^5} = 2Dk_0 + C\ \{(\sigma^{-3} + 3\sigma^{-5}) \sinh\sigma - 3\sigma^{-4}\cosh\sigma\}, \tag{16.7}$$

$$\frac{3A}{R^4} - \frac{5B}{R^6} = \frac{C}{\sqrt{k_0}}\{(-6\sigma^{-4} + 15\sigma^{-6}) \sinh\sigma + (\sigma^{-3} + 15\sigma^{-5}) \cosh\sigma\}, \tag{16.8}$$

$$-\frac{12A}{R^5} + \frac{30B}{R^7} = \frac{C}{k_0}\{(\sigma^{-3} + 39\sigma^{-5} + 90\sigma^{-7}) \sinh\sigma - (9\sigma^{-4} + 90\sigma^{-6}) \cosh\sigma\}, \tag{16.9}$$

where c was defined by (10.13). Solving the four unknown constants one finds

$$A = \frac{5}{2} R^3 G_1(\sigma)\ \{1 + \frac{10}{\sigma^2} G_1(\sigma)\}^{-1}, \tag{16.10}$$

where

$$G_1(\sigma) = 1 + \frac{3}{\sigma^2} - \frac{3}{\sigma} \operatorname{cotgh} \sigma . \tag{16.11}$$

Note that this function also plays a role in the rotational diffusion coefficient of a porous sphere, eq. (10.17).

With (15.24) this gives for the relative increase of the viscosity of a dilute solution of uniformly porous spheres

$$\frac{\eta-\eta_0}{\eta_0} = \frac{10}{3}\pi R^3 n_p G_1(\sigma) \{1 + \frac{10}{\sigma^2} G_1(\sigma)\}^{-1} . \tag{16.12}$$

Just as in the case of the translational diffusion coefficient this result was first published by Debye and Bueche (1948), but the first derivation was given by Felderhof (1975b). Introducing the fraction of the volume occupied by the spheres

$$\Omega = \frac{4}{3}\pi R^3 n_p \tag{16.13}$$

one can write (16.12) in the more popularized form

$$\frac{\eta-\eta_0}{\eta_0} = \frac{5}{2}\Omega G_1(\sigma) \{1 + \frac{10}{\sigma^2} G_1(\sigma)\}^{-1} . \tag{16.14}$$

In the limit $\sigma \to \infty$ the spheres become impermeable. As $G_1(\infty) = 1$ this gives

$$\frac{\eta-\eta_0}{\eta_0} = \frac{5}{2}\Omega \qquad \text{(impermeable spheres)}, \tag{16.15}$$

the result originally found by Einstein (1906, 1911, 1956). In the free draining regime one finds

$$\frac{\eta-\eta_0}{\eta_0} = \frac{2}{15}\pi R^3 n_p \sigma^2, \qquad (\sigma \ll 1). \tag{16.16}$$

In table V we give the values of the correction factor $G_1(\sigma) \{1 + \frac{10}{\sigma^2} G_1(\sigma)\}^{-1}$ due to the finite permeability of the spheres.

σ	$G_1(\sigma)\,\{1+\frac{10}{\sigma^2}\,G_1(\sigma)\}^{-1}$
0	0
1	0.0378
2	0.1306
3	0.2406
4	0.3432
5	0.4301
6	0.5020
7	0.5603
8	0.6081
9	0.6475
10	0.6803
∞	1

Table V. Correction factor for the relative increase in the viscosity due to a finite permeability of the suspended spheres, eq. (16.12).

17. VISCOSITY OF GAUSSIAN COILS

In this section we study the case of Gaussian coils

$$k(r) = K \exp(Qr^2), \tag{17.1}$$

which also formed the subject of sections 13 and 14. The treatment given here follows Wiegel and Mijnlieff (1977b). The appropriate dimensionless variables are

$$x = r\sqrt{Q}, \tag{17.2}$$

$$f(x) = \phi(r), \tag{17.3}$$

$$g(x) = \chi(r). \tag{17.4}$$

In terms of these variables the equations (15.7,8) become

$$\frac{d^2f}{dx^2} + \frac{6}{x}\frac{df}{dx} - \alpha e^{-x^2} f + \frac{1}{x}\frac{dg}{dx} = 0, \tag{17.5}$$

$$\frac{d^2g}{dx^2} + \frac{2}{x}\frac{dg}{dx} - \frac{6}{x^2} g + 2\alpha x^2 e^{-x^2} f = 0, \tag{17.6}$$

where α was defined by (14.7). The boundary conditions are: $f(0)$ finite, $g(0) = 0$, $f(\infty) = 1$ and $\lim_{x\to\infty} \frac{g(x)}{x} = 0$. An analysis similar to the one in section 14 shows that $g'(0) = 0$ and that $f'(0) = 0$.

A numerical solution of the equations can be found in a straightforward way. Outside the coil (15.13,14) give

$$g(x) = \frac{2a}{x^3}, \tag{17.7}$$

$$f(x) = 1 - \frac{a}{x^3} + \frac{b}{x^5}. \tag{17.8}$$

The two constants a and b are adjusted till the solution satisfies the boundary conditions $g(0) = 0$, $f'(0) = 0$. Denoting the correct value of a by $\alpha\Phi(\alpha)$ one finds

$$A = Q^{-\frac{3}{2}} a = K^{-1} Q^{-\frac{5}{2}} \Phi(\alpha). \tag{17.9}$$

The viscosity follows from the last equation and (15.24)

$$\frac{\eta-\eta_0}{\eta_0} = \frac{4}{3}\pi n_p K^{-1} Q^{-\frac{5}{2}} \Phi(\alpha). \tag{17.10}$$

The values of $\Phi(\alpha)$ have been calculated in this way for α up to 17; the results are listed in table IV.

For the record, it should be pointed out that in chemical physics the change in the viscosity is usually expressed in terms of the intrinsic viscosity $[\eta]$ which is defined as the relative increase in the viscosity divided by the mass density of macromolecules in the mixture. Denoting the mass of a single coil by m one finds

$$[\eta] = \frac{4}{3}\pi\, m^{-1}\, K^{-1}\, Q^{-\frac{5}{2}}\, \Phi(\alpha). \tag{17.11}$$

This result was given by Wiegel and Mijnlieff (1977a,b); its comparison with the experimental data is discussed briefly in the next section.

18. COMPARISON WITH EXPERIMENTAL DATA

The theory of the translational friction coefficient and intrinsic viscosity of Gaussian coils, developed in sections 14 and 17, has been compared by Mijnlieff and Wiegel (1978) with experimental data on two polymer solvent pairs. These two systems were (A) poly-α-methylstyrene in cyclohexane at 35.5^0 C; (B) poly-α-methylstyrene in toluene at 25^0 C. Note that cyclohexane is a poor solvent, but toluene a good solvent for poly-α-methylstyrene. The physical consequences of this fact have been discussed at the end of section 4. System A is at its theta temperature, but system B is far from theta conditions. Hence the two systems represent very different situations.

The "radius of gyration" ρ_g and the molecular mass m of the macromolecule follow from sedimentation and light scattering experiments. Using ρ_g and m in equations (13.16,17) one finds Q and the mass density $\rho_1(0)$ in the center of the coil. Eq. (13.15) then leads to the value of K.

The theoretical values of f_T and $[\eta]$ thus calculated turn out to be in satisfactory agreement with the experimental data. Calculated and directly measured values agree within 5% tot 15%. This is satisfactory in view of the fact that no a priori information on f_T or $[\eta]$ has been used, and that no adjustable parameters occur in the theory. It should be pointed out also that the uncertainty in the experimental values of these transport coefficients is of the order of 10%.

19. SOME GENERAL PROPERTIES OF CELL MEMBRANES

The flow of a viscous fluid through a porous macromolecular system also plays a role in the biophysics of the cell membrane. One of the most important examples is the case in which the porous medium consists of a patch of cross-linked immunoglobulins through and around which exists a lateral flow of lipids. Before embarking upon any calculations we shall discuss the two systems involved (the lipid bilayer and the cross-linked immunoglobulins) in a qualitative fashion.

It is now generally believed that most biological membranes consist of a lipid bilayer in which a large number of proteins are embedded (Singer and Nicolson, 1972). Upon variation of the temperature or of some other appropriate variable the bilayer can be "pushed" into one of several phases. The current theories of these phases and the corresponding phase transitions have recently been reviewed in detail by Wiegel and Kox (1979). The interested reader is referred to this paper and the references cited therein. At physiological conditions the bilayer is in a fluid phase.

The lipid bilayer has a thickness

$$h \cong 40\ \text{Å} \tag{19.1}$$

and a viscosity

$$\eta \cong 2\,\mathrm{g\,cm^{-1}\,s^{-1}} \tag{19.2}$$

at 25° C (Prives and Shinitzky, 1977). Note that this viscosity is quite large as compared to the viscosity η' of the fluid on either side of the membrane; η' is of the order of the viscosity of water

$$\eta \cong 0.01\,\mathrm{g\,cm^{-1}\,s^{-1}} \tag{19.3}$$

at 25° C (Weast, 1974). The lipids in each of the two monolayers can be represented roughly by hard disks of a radius

$$a \cong 4\ \text{Å} \tag{19.4}$$

and a height approximately equal to half the thickness of the membrane. As lipids which are opposite to each other in the two monolayers interact strongly with each other through entanglements of the ends of their hydrocarbon tails we can imagine the membrane to consist of hard rods (of length h), which are constrained to move in the plane of the membrane and which are always oriented perpendiculary to its surface. This model was first used by Huang (1973).

If the biological cell under consideration is a lymphocyte the lipid bilayer will be interrupted by immunoglobulin molecules. A detailed discussion of these proteins can be found in De Lisi (1976) and in the references quoted there. Mathematical immunology also forms the subject of a recent textbook by Bell, Perelson and Pimbley (1978), where the reader will find much related material. For our purposes the following three observations are crucial: 1. The immunoglobulin molecules (specifically the IgG molecules) are large Y shaped proteins which are embedded in the lymphocyte membrane with their Fc fragment, i.e. with the "stem" of the Y. 2. These immunoglobulins are often cross-linked in a patch consisting of hundreds or thousands of them. The cross-linking occurs through binding of other compounds (antigens) outside the membrane to the two "prongs" of the Y. 3. As the linear dimension of an IgG molecule is of the order of 100 Å the distance between the stems of neighboring immunoglobulins in a patch is of order of 100 Å too. As this is large compared to the size (19.4) of a lipid the patches are permeable for lateral flow in the plane of the lipid bilayer.

20. THE BRETSCHER FLOW HYPOTHESIS

An interesting addition to the concept of the fluid membrane is due to Bretscher (1976). This author assumed that "new" lipids are inserted at certain sites of the membrane and that "old" lipids are removed through endocytosis at one specific site (or a few specific sites). This continuous recycling of the lipids leads to a continuous oriented lateral flow in the plasma membrane. As a consequence of this flow all objects in the membrane (isolated proteins or complexes of cross-linked proteins) are subject to a drag force which is directed towards the specific site where lipids are removed. This sweeping action of membrane flow is counteracted by the Brownian motion and the resulting distribution of the objects in the membrane is the result of the competition between these two effects.

In the following sections we pursue some of the more important consequences of the Bretscher flow hypothesis in a quantitative way. We shall base our considerations on a model with the following simplifying features: 1. The cell membrane is a sphere of constant radius R. Typically R is of the order 10^4 to 10^5 Å. 2. The system is in the stationary state. 3. Lipids are removed at only one specific site. 4. Lipids are inserted at random positions. 5. The interactions between the different diffusing objects are neglected. The calculations in this and the following sections follow a paper by Wiegel (1979a).

It is convenient to use polar coordinates θ,ϕ on the surface of the cell in such a way that the specific point at which lipids are removed from the membrane through endocytosis corresponds to the "North Pole" $\theta = 0$. New lipids are inserted in the membrane at random positions in such a way that in the stationary state a fraction α of the total area is renewed per unit of time. Hence α roughly measures the intensity of the lipid metabolism of the cell. The local velocity $v(\theta)$ is directed tangentially to the membrane, in the direction of decreasing values of θ. The surface element of the sphere equals

$$d^2S = 2\pi R^2 \sin\theta \, d\theta. \tag{20.1}$$

As the lipid bilayer flows as an incompressible fluid a consideration of conservation of mass gives

$$v(\theta) = \alpha R(1+\cos\theta)/\sin\theta . \tag{20.2}$$

The average magnitude of this velocity over the entire surface of the cell is

$$\langle v\rangle = (4\pi R^2)^{-1} \int v(\theta) d^2S$$

$$= \tfrac{1}{2}\pi\alpha R . \tag{20.3}$$

Bretscher (1976) gives $\langle v\rangle \cong 5 \times 10^{-6}$ cm s^{-1} as a typical average velocity.

If one generalizes the model under consideration - for example by permitting two or more lipid "sinks" - the Bretscher hypothesis leads to problems related to two-dimensional incompressible hydrodynamics on a spherical surface. This problem has recently been considered by Buas (1977) in some detail.

21. EQUILIBRIUM DISTRIBUTION OF DIFFUSING PARTICLES

Suppose there are N particles embedded in the cell membrane, each with the same translational diffusion coefficient D_T. The particles have no interactions; for example: no coagulation occurs. What is their equilibrium distribution over the membrane?

In order to answer this question first consider the situation in which the particles under consideration are embedded in a membrane which flows with some time independent velocity $\vec{v}_0(\vec{r})$. If a tagged particle has the instantaneous velocity $\vec{v}(t)$ its motion can be described by the Langevin equation

$$m\frac{d\vec{v}}{dt} = -f_T(\vec{v}-\vec{v}_0) + \vec{\varepsilon}(t). \tag{21.1}$$

In this equation m denotes the mass of the particle, f_T its translational friction coefficient and $\vec{\varepsilon}(t)$ a rapidly fluctuating stochastic force in the plane of the membrane (compare the detailed discussion of the Langevin equation in Chandrasekhar (1943), reprinted in Wax (1954)). The last equation can be written also in the form

$$m\frac{d\vec{v}}{dt} = f_T\,\vec{v}_0 - f_T\,\vec{v} + \vec{\varepsilon}(t). \tag{21.2}$$

In this form, however, the equation is identical to the equation which would describe the movement of the same particle immersed in a fluid at rest, but subject to an external force equal to $f_T\vec{v}_0$. The potential $\Phi(\vec{r})$ corresponding to this external force is given by the line integral

$$\Phi(\vec{r}) = -f_T\int \vec{v}_0(\vec{r}\,')\cdot d\vec{r}\,' \tag{21.3}$$

taken along a contour which ends in the point $\vec{r}$. According to statistical mechanics the local surface density $\rho(\vec{r})$ of particles (the number of particles per unit area) is proportional to the Boltzmann factor

$$\rho(\vec{r}) \sim \exp\{-\Phi(\vec{r})/k_BT\}. \tag{21.4}$$

In the special case (20.2) the potential is given by

$$\begin{aligned}\Phi(\theta) &= f_T\int^{\theta} \alpha R^2\,\frac{1+\cos\theta'}{\sin\theta'}\,d\theta' \\ &= 2\alpha R^2 f_T \,\ln(\sin\tfrac{1}{2}\theta) + \text{constant}.\end{aligned} \tag{21.5}$$

Combination of the last two equations gives the number density $\rho(\theta)$ of the diffusing particles

$$\rho(\theta) = \rho(\pi)\,(\sin\tfrac{\theta}{2})^{-2\alpha R^2/D_T}, \tag{21.6}$$

where $\rho(\pi)$ denotes the number density of diffusing particles at the "South Pole" and where the Einstein relation (8.1) was used to express the friction coefficient in terms of the diffusion coefficient.

The calculation leading to this result shows that the equilibrium distribution of diffusing particles can be derived explicitly provided the interactions between the particles can be neglected. In some cases of biophysical interest, as in the case of patches of coagulating immunoglobulins, the interactions between the diffusing particles are essential. Aggregate formation on lymphocyte membranes in the absence of diffusion- and convection processes has been studied by Perelson and De Lisi (1975) and by De Lisi and Perelson (1976); the theory has been reviewed by Perelson (1979). Aggregate formation in the presence of diffusion- and convection processes forms the subject of current work by Perelson and Wiegel (1979).

22. CAP FORMATION

According to the preceeding section the diffusing particles are distributed over the cell surface with a density which increases from a finite value $\rho(\pi)$ at the South Pole to infinity at the North Pole. The actual divergence of $\rho(\theta)$ if $\theta \downarrow 0$ is an artefact of a model in which the mutual interactions between the diffusing particles are neglected. If the particles are assumed to have a finite diameter, or if endocytosis is assumed to occur in a finite neighborhood or $\theta = 0$, the divergence in $\rho(\theta)$ should be replaced by a sharp peak of finite height. To the author's knowledge no results have been published for particles with interactions, as for example hard disks diffusing in the surface of a sphere.

In any case the result (21.6) can be accepted as a rough approximation to the exact distribution. The density (21.6) has the remarkable property that its integral over the whole membrane surface will diverge if $2\alpha R^2/D_T \geq 2$. This divergence implies that a Brownian particle, if picked at random, will be found in an infinitesimal vicinity of $\theta = 0$ with probability unity. The resulting clustering phenomenon can be interpreted to mean that the Brownian particles form a cap at $\theta = 0$. This leads to a quantitative criterion for cap formation

$$\text{(cap formation)} \quad \leftrightarrow (\alpha R^2 \geq D_T). \qquad (22.1)$$

From a biophysical point of view this criterion implies that the cell can form a cap from globular proteins or permeable patches of cross-linked immunoglobulins by increasing its lipid metabolism (recall that $4\pi\alpha R^2$ equals the total area of the cell membrane which gets renewed per unit of time). The same can be accomplished by a decrease of the diffusion coefficient, which itself results from an increase in membrane viscosity. We can proceed now to the theory of the translational and rotational diffusion coefficients of small objects immersed in a viscous lipid membrane.

The mechanism suggested in this section by no means represents the only theory of cap formation. A discussion of the wealth of observations related to cap formation is beyond the limited scope of these lectures. The interested reader is refered to the review paper by Schreiner and Unanue (1976).

23. GENERAL THEORY OF THE ROTATIONAL DIFFUSION COEFFICIENT

In this section we consider a porous cylinder of infinite length and uniform permeability immersed in an incompressible fluid (viscosity η) and rotating with a constant angular velocity ω_0 around its axis. We calculate the rotational friction coefficient f_R, which is defined as the ratio τ_3/ω_0, where $\vec{\tau}$ denotes the total torque of the forces per unit of length which the cylinder exerts on the fluid. The results of this and the next section were published by Wiegel (1979a,b). The situation corresponds to the $h = \infty$ limit of figure 3 which shows a membrane of thickness h and viscosity η, embedded on both sides in a fluid of much lower viscosity η'. The porous cylinder (radius a) is a model for a permeable patch of cross-linked immunoglobulins. The problem is determined by the dimensionless parameter $a\eta'/h\eta$. In this section we consider only the limiting case in which $a\eta'/h\eta = 0$; correction terms (which are small because $\eta'/\eta \simeq 0.005$) should not be difficult to calculate.

Consider a cartesian system of coordinates (x,y,z) with the z-axis along the axis of the cylinder (we shall also use cylindrical coordinates r,ϕ,z). Let $\vec{V}$ and P denote the average local velocity and pressure of the fluid and $\vec{U}$ the local velocity of the cylinder. The fluid flow has to be solved from the fundamental equation (5.5)

$$-\vec{\nabla} P + \eta \Delta \vec{V} + \frac{\eta}{k} (\vec{U} - \vec{V}) = 0 \tag{23.1}$$

together with the incompressibility equation (5.4)

$$\mathrm{div}\, \vec{V} = 0. \tag{23.2}$$

The x- and y components of $\vec{U}$ are

$$U_1 = -\omega_0 r \sin\phi , \tag{23.3}$$

$$U_2 = +\omega_0 r \cos\phi . \tag{23.4}$$

For the pressure and the velocity we make the Ansatz

$$P = \text{constant}, \tag{23.5}$$

$$V_1 = - V(r) \sin\phi , \tag{23.6}$$

$$V_2 = + V(r) \cos\phi , \tag{23.7}$$

where V(r) denotes an unknown function - the magnitude of the velocity - which has cylindrical symmetry. Upon substitution of (23.5-7) into (23.1,2) one finds that all equations are satisfied provided V(r) is the solution of the ordinary differential equation

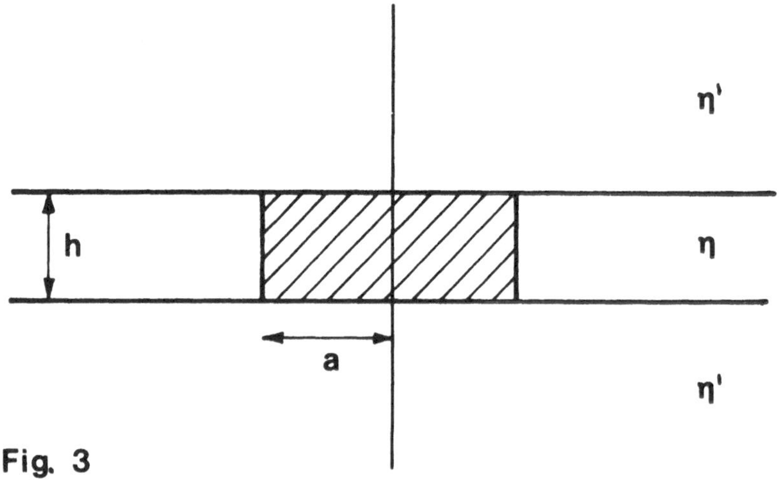

Fig. 3

Basic geometry for the hydrodynamics of the cell membrane (discussion in section 23).

$$V'' + \frac{1}{r} V' - \frac{1}{r^2} V = \frac{V}{k} - \frac{\omega_0 r}{k} . \tag{23.8}$$

The prime denotes differentiation with respect to the radius r. The boundary conditions are $|V(0)| < \infty$ and $V(\infty) = 0$. In the next section we solve this equation for the important case of the uniform cylinder.

24. ROTATIONAL DIFFUSION COEFFICIENT OF A UNIFORM DISK

For the uniform disk of radius a one has

$$k(r) = k_0 \qquad (r < a), \qquad (24.1)$$

$$= \infty \qquad (r > a). \qquad (24.2)$$

Substituting into the last equation of the previous section one finds a differential equation with the solution

$$V(r) = \frac{A}{r} \qquad (r > a), \qquad (24.3)$$

$$= \omega_0 r + B\, I_1(r/\sqrt{k_0}) \qquad (0 < r < a). \qquad (24.4)$$

In these and the following equations the I_ν denote the modified Bessel functions.

The constants A and B follow from the condition that V(r) and V'(r) should be continuous at $r = a$ (compare the discussion in section 10). This gives

$$A = \omega_0 a^2 \frac{I_2(\sigma)}{I_0(\sigma)}, \qquad (24.5)$$

$$B = \frac{-2\omega_0 a}{\sigma I_0(\sigma)}. \qquad (24.6)$$

Here

$$\sigma = \frac{a}{\sqrt{k_0}}, \qquad (24.7)$$

analogous to (10.13).

The torque of the forces which the cylinder exerts on the fluid, per unit of length, has a z-component equal to

$$\tau_3 = -\frac{2\pi\eta B}{k_0} \int_0^a r^2 I_1\left(\frac{r}{\sqrt{k_0}}\right) dr = 4\pi\eta\omega_0 a^2 \frac{I_2(\sigma)}{I_0(\sigma)}. \qquad (24.8)$$

Hence the rotational friction coefficient for a cylinder of length h is given by

$$f_R = 4\pi\eta h a^2 \frac{I_2(\sigma)}{I_0(\sigma)}. \qquad (24.9)$$

Finally, the Einstein relation (8.2) gives for the rotational diffusion coefficient

$$D_R = \frac{k_B T}{4\pi\eta h a^2} \frac{I_0(\sigma)}{I_2(\sigma)}. \qquad (24.10)$$

The application of this result to the rotational diffusion of globular proteins or patches of immunoglobulins in the cell membrane is the subject of section 27.

In the limit $\sigma \to \infty$ the cylinder becomes impermeable and one recovers the trivial result

$$f_R(\infty) = 4\pi\eta h a^2 \qquad (24.11)$$

which holds for a hard cylinder. If, on the other hand, $\sigma \ll 1$ the expression (24.9) simplifies to

$$f_R \cong \tfrac{1}{2}\pi\eta h a^2 \ \sigma^2, \qquad (\sigma \ll 1); \qquad (24.12)$$

here we used eq. 9.6.7 of Abramowitz and Stegun (1970). This also can be derived independently with perturbation theory as $\sigma \ll 1$ corresponds to the free draining regime. Using table 9.8 of Abramowitz and Stegun (1970) we have tabulated the correction factor $I_2(\sigma)/I_0(\sigma)$ due to the finite permeability of the cylinder (table V).

σ	$\frac{I_2(\sigma)}{I_0(\sigma)}$
0	0
1	0.1072
2	0.3022
3	0.4600
4	0.5682
5	0.6426
6	0.6958
7	0.7355
8	0.7662
9	0.7905
10	0.8103
∞	1

Table V. Correction factor in the rotational friction coefficient of a porous cylinder, eq. (24.9), due to a finite permeability of the cylinder.

25. GENERAL THEORY OF THE TRANSLATIONAL DIFFUSION COEFFICIENT

In order to calculate the translational friction coefficient we must consider the following problem. A cylinder (radius a and height h) with given permeability k(r) is in a state of uniform translation with respect to a viscous fluid (viscosity η). This fluid forms a layer of constant thickness h in which the cylinder is constrained to move. The fluid sheet is embedded on both sides in another fluid of a much lower viscosity (η'). The situation is drawn in figure 3. One wants to calculate the translational friction coefficient f_T.

For an impermeable cylinder in the limit $(a\eta'/h\eta) \to 0$ this problem has no solution due to the Stokes paradox (compare the discussion in §20 of Landau and Lifshitz (1959) and in Buas (1977)). It follows from the analysis in this and the next section that there also is no solution in the limit $(a\eta'/h\eta) \to 0$ for any finite value of the permeability, i.e. even when the cylinder has a certain porosity the Stokes paradox still obtains.

In the case $a\eta'/h\eta > 0$, however, the situation is quite different. It was pointed out by Saffman and Delbrück (1975) for the case of an impermeable cylinder, that the problem does have a solution as long as $a\eta'/h\eta > 0$. Saffman (1976) has presented the details of this calculation using singular perturbation theory (Van Dyke, 1975). In this section and the next we consider the case in which the cylinder has finite permeability, and we calculate the leading term in f_T. The results have been published by Wiegel (1979a,b,c). As in section 24 we use a Cartesian system of coordinates (x,y,z) with the z-axis along the axis of the cylinder, and later on also cylindrical coordinates (r,ϕ,z). The sheet of fluid with viscosity η is located at $-h < z < 0$, the other fluid with much lower viscosity η' is located at $z < -h$ and at $z > 0$.

First, we consider the pressure (p) and velocity $\vec{v} = (v_1, v_2, v_3)$ of the fluid in the half space $z > 0$. Putting the cylinder at rest we have to solve the linearized, time - independent Navier - Stokes equation

$$-\vec{\nabla} p + \eta' \Delta \vec{v} = 0 \tag{25.1}$$

and the incompressibility equation

$$\mathrm{div}\, \vec{v} = 0 \tag{25.2}$$

under the boundary condition that at large distances from the z-axis $v_1 \to -v_0$, $v_2 \to 0, v_3 \to 0$. For pressure and velocity we use the Ansatz

$$v_1 = -v_0 + \{s(r,z) + t(r,z)\} \cos^2\phi - t(r,z), \tag{25.3}$$

$$v_2 = \{s(r,z) + t(r,z)\} \cos\phi \sin\phi , \tag{25.4}$$

$$v_3 = 0, \tag{25.5}$$

$$p = \text{constant}, \tag{25.6}$$

where s and t denote unknown functions with cylindrical symmetry which vanish for $r \to \infty$ or $z \to \infty$. Physically $(\delta v)_r \equiv s(r,z) \cos\phi$ equals the component in the direction of increasing r- values of the perturbation of the asymptotic velocity field, and $(\delta v)_\phi \equiv t(r,z) \sin\phi$ equals the component in the direction of increasing ϕ -values. Substitution into the continuity equation (25.2) immediately leads to

$$t = -s - r \frac{\partial s}{\partial r} \tag{25.7}$$

which relation reduces the number of unknown functions to one. Substituting the Ansatz into (25.1) and eliminating t(r,z) with the last relation we find

$$\left(\frac{\partial^2}{\partial r^2} + \frac{3}{r} \frac{\partial}{\partial r} + \frac{\partial^2}{\partial z^2}\right) s(r,z) = 0. \tag{25.8}$$

The general solution which is finite at $r = 0$ and vanishes if $r \to \infty$ or $z \to \infty$ equals

$$s(r,z) = \int_0^\infty f(k) \frac{J_1(kr)}{kr} \exp(-kz) dk, \tag{25.9}$$

where the J_ν denote the Bessel functions of the first kind and where f(k) still has to be determined from the boundary condition at the fluid-fluid interface at $z = 0$.

Second, we consider the fluid in the sheet $-h < z < 0$. In applications to the cell membrane this fluid consists of lipids directed along the z-axis. Hence the flow in the sheet is truly two-dimensional (compare the discussion in section 19).

In the presence of a porous cylinder we describe the fluid in the sheet in terms of the (space-time) average pressure P and the average local velocity $\vec{V}$. If the two fluids did not interact at their interfaces $z = 0$ and $z = -h$ the flow would follow from the Debye-Brinkman-Bueche equation (5.5)

$$-\vec{\nabla} P + \eta \Delta \vec{V} - \frac{\eta}{k} \vec{V} = 0, \tag{25.10}$$

where $k = k(r)$ denotes the permeability which is assumed to have cylindrical symmetry.

In the presence of interaction between the two fluids at their interfaces the fluid located at $z > 0$ exerts a force $\vec{\sigma}$ (per unit area) on the fluid in the sheet which is given by

$$\vec{\sigma} = \eta' \left(\frac{\partial \vec{v}}{\partial z}\right)_{z=0}; \tag{25.11}$$

an identical force derives from the second interface at $z = -h$. Hence P and $\vec{V}$ have to be solved from

$$-\vec{\nabla} P + \eta \Delta \vec{V} - \frac{\eta}{k} \vec{V} + \frac{2\vec{\sigma}}{h} = 0, \qquad (25.12)$$

$$\text{div}\, \vec{V} = 0. \qquad (25.13)$$

The shear force $\vec{\sigma}$ should of course not be confused with the dimensionless parameter σ which was used in (24.7) and (10.13).

For pressure and velocity in the sheet we make the Ansatz

$$V_1 = -v_0 + \{S(r) + T\ (r)\}\ \cos^2\phi - T(r), \qquad (25.14)$$

$$V_2 = \{S(r) + T\ (r)\}\ \cos\phi \sin\phi\ , \qquad (25.15)$$

$$P = \eta \Pi\ (r)\ \cos\phi\ , \qquad (25.16)$$

where S, T and Π denote unknown functions with cylindrical symmetry which vanish in the limit $r \to \infty$. This Ansatz is similar to (25.3-6). Substitution into the continuity equation (25.13) leads to

$$T = -S - r\,\frac{dS}{dr}\ . \qquad (25.17)$$

Substitution of the Ansatz into (25.12) leads to two ecuations which, after some straightforward but tedious algebra, can be rewritten in the form

$$-\Pi' + S'' + \frac{3}{r} S' - \frac{S}{k} + \frac{v_0}{k} + \frac{2\eta'}{h\eta} \left(\frac{\partial s}{\partial z}\right)_{z=0} = 0, \qquad (25.18)$$

$$-r\Pi'' - \Pi' + \frac{\Pi}{r} - r(k^{-1})'S + r(k^{-1})'v_0 = 0, \qquad (25.19)$$

where the prime denotes differentiation with respect to r.

These equations have to be solved under the boundary conditions: (i) $\Pi(0)$ and $S(0)$ should be finite; (ii) $\Pi(\infty) = S(\infty) = 0$; (iii) $s(r,0) = S(r)$, which condition expresses the continuity of the flow field in the sheet with the flow fields outside the sheet at the interfaces at $z = 0$ and $z = -h$. The last term on the left hand side of (25.18) is determined uniquely by the function $S(r)$ through (25.9) and boundary condition (iii). In the next section we solve these equations for the important case of a cylinder of constant permeability. Even in this simple case the solution can be found only in the asymptotic regime $a\eta'/h\eta \ll 1$.

26. TRANSLATIONAL DIFFUSION COEFFICIENT OF A UNIFORM DISK

The uniform disk was defined in equations (24.1-2). With a discontinuity in $k(r)$ at $r = a$ eq. (25.12) shows that pressure, velocity and all first derivatives of the velocity should be continuous at the surface of the disk, but that the derivative of the pressure and/or the second derivatives of the velocity may be discontinuous. Equation (25.19) now reads

$$\Pi'' + \frac{1}{r}\Pi' - \frac{1}{r^2}\Pi = 0, \tag{26.1}$$

with the solution

$$\Pi(r) = L\,r \qquad (0 < r < a), \tag{26.2}$$

$$= \frac{La^2}{r} \qquad (a < r), \tag{26.3}$$

where L is a constant. Equations (25.9) and (25.18) together with boundary condition (iii) of section 25 now give

$$-L + S'' + \frac{3}{r}S' - \frac{S}{k_0} + \frac{v_0}{k_0} - \frac{2\eta'}{h\eta r}\int_0^\infty f(k)J_1(kr)dk = 0 \qquad (0 < r < a), \tag{26.4}$$

$$\frac{La^2}{r^2} + S'' + \frac{3}{r}S' - \frac{2\eta'}{h\eta r}\int_0^\infty f(k)J_1(kr)dk = 0 \qquad (a < r), \tag{26.5}$$

$$S(r) = \int_0^\infty f(k)\frac{J_1(kr)}{kr}dk. \tag{26.6}$$

It is easy to verify that these equations have no solution which satisfies the boundary conditions at $r = 0$ and $r = \infty$ in the limit $(a\eta'/h\eta) \to 0$. Hence the Stokes paradox also holds for a porous cylinder of infinite length. For $0 < \frac{a\eta'}{h\eta} \ll 1$, however, the equations do have a solution which can be found with singular perturbation theory (compare Van Dyke, 1975).

For $r \gg a$ the force density $-\frac{\eta}{k}\vec{V}$ in (25.12) can be replaced by a sharply peaked force density $\vec{F}$ in the origin, with x- and y-components

$$F_1 = \frac{F}{\pi h r}\,\delta(r), \tag{26.7a}$$

$$F_2 = 0. \tag{26.7b}$$

Here F denotes the magnitude of the total force which the cylinder exerts on the fluid; we also made use of the fact that the delta function $\delta(r)$ is an even function. The asymptotic behavior of the flow field for $r \gg a$ can thus be solved from the much simpler equation

$$-\vec{\nabla} P + \eta \Delta \vec{V} + \vec{F} + \frac{2\vec{\sigma}}{h} = 0. \tag{26.8}$$

It is straightforward, but somewhat tedious, to solve this equation using the methods of the previous section. Instead of (25.18,19) one finds the equations

$$-\Pi' + S'' + \frac{3}{r} S' + \frac{F\delta(r)}{\pi\eta h r} + \frac{2\eta'}{h\eta} \left(\frac{\partial s}{\partial z}\right)_{z=0} = 0, \tag{26.9}$$

$$- r \Pi'' - \Pi' + \frac{\Pi}{r} + \frac{F}{\pi\eta h} r \left(\frac{\delta(r)}{r}\right)' = 0. \tag{26.10}$$

The last equation can be integrated immediately; using the boundary condition on Π at infinity we find

$$-\Pi' + \frac{F\delta(r)}{\pi\eta h r} = \frac{\Pi}{r}. \tag{26.11}$$

The solution of this equation which has the appropriate behavior near $r = 0$ is given by

$$\Pi(r) \cong \frac{F}{2\pi\eta h r} \qquad (r \gg a). \tag{26.12}$$

Note that this asymptotic behavior of the pressure has the same form as the exact result (26.3).

Upon substitution of the last two equations (26.9) takes the form

$$S'' + \frac{3}{r} S' + \frac{F}{2\pi\eta h r^2} + \frac{2\eta'}{h\eta} \left(\frac{\partial s}{\partial z}\right)_{z=0} = 0. \tag{26.13}$$

Taking the Hankel transform and using (25.9) and (26.6) we find the solution in the form of a definite integral

$$S(r) \cong \frac{F}{2\pi\eta h} \int_0^\infty (\xi^2 + 2\eta' r\xi/h\eta)^{-1} J_1(\xi) d\xi \qquad (r \gg a). \tag{26.14}$$

This will be called the outer asymptotic expansion. This solution also obeys the proper boundary condition for $r \to \infty$. Using the recurrence relations for the Bessel functions (Abramowitz and Stegun, 1970, eq. 9.1.27) the integral can be written as the sum of two terms

$$\int_0^\infty (\xi^2 + 2\eta' r/h\eta)^{-1} J_1(\xi) d\xi = \frac{1}{2} \int_0^\infty (\xi + 2\eta' r/h\eta)^{-1} J_0(\xi) d\xi + \frac{1}{2} \int_0^\infty (\xi + 2\eta' r/h\eta)^{-1} J_2(\xi) d\xi. \tag{26.15}$$

The second integral on the right hand side converges even for $r = 0$, and has the value (Gradshteyn and Ryzhik, 1965, eq. 6.561.17)

$$\tfrac{1}{2} \int_0^\infty (\xi + 2\eta' r/h\eta)^{-1} J_2(\xi) d\xi \cong \tfrac{1}{4} \qquad (r \ll \frac{h\eta}{\eta'}). \qquad (26.16)$$

The first integral on the right hand side of (26.15) has the value (Gradshteyn and Ryzhik, 1965, eq. 6.562.2)

$$\tfrac{1}{2} \int_0^\infty (\xi + 2\eta' r/h\eta)^{-1} J_0(\xi) d\xi = \frac{\pi}{4} [H_0(2\eta' r/h\eta) - N_0(2\eta' r/h\eta)] , \qquad (26.17)$$

where the N_ν denote the Neumann functions and the H_ν the Struve functions. Using the asymptotic expansions of these functions for small arguments, as given by Gradshteyn and Ryzhik (1965) eq. 8.550 and 8.403.2, and combining the last four equations we find for the inner limit of the outer asymptotic expansion of the flow

$$S(r) \cong \frac{F}{4\pi\eta h} \{ \tfrac{1}{2} - \gamma + \ln (\frac{h\eta}{\eta' r}) \} \qquad (a \ll r \ll \frac{h\eta}{\eta'}), \qquad (26.18)$$

where $\gamma = 0.5772$ denotes Euler's constant. The constant F will be determined shortly.

For $r \ll h\eta/\eta'$ the last term on the left hand side of (26.4) and (26.5) can be neglected with respect to the terms $S'' + \frac{3}{r} S'$. Hence the inner asymptotic expansion can be solved from the equations

$$-L + S'' + \frac{3}{r} S' - \frac{S}{k_0} - \frac{v_0}{k_0} = 0 \qquad (0 < r < a), \qquad (26.19)$$

$$\frac{La^2}{r^2} + S'' + \frac{3}{r} S' = 0 \qquad (a < r). \qquad (26.20)$$

The solution of these equations which is finite for $r \downarrow 0$ (the inner asymptotic expansion) equals

$$S(r) = v_0 - L k_0 + A \frac{I_1(r/\sqrt{k_0})}{r/\sqrt{k_0}} \qquad (0 < r < a), \qquad (26.21)$$

$$S(r) = \frac{\alpha}{r^2} + \beta - \tfrac{1}{2} L a^2 \ln r \qquad (a < r), \qquad (26.22)$$

where A, α and β are constants.

As was noted in the beginning of this section the pressure, velocity and all first derivatives of the velocity should be continuous at $r = a$. With (25.14-17) this implies that S, S' and S'' should be continuous at $r = a$. With the explicit results (26.21) and (26.22) this leads to three equations from which the three constants α,

β and A can be solved as functions of L. The solution of these equations is given by

$$A = -\frac{La^2}{\sigma I_1(\sigma)}, \tag{26.23}$$

$$\frac{2\alpha}{a^2} = -L\,a^2\,\{\tfrac{1}{2} + \frac{2}{\sigma^2} - \frac{I_0(\sigma)}{\sigma I_1(\sigma)}\}, \tag{26.24}$$

$$\beta = v_0 + L\,a^2\,\{-\frac{1}{\sigma^2} + \tfrac{1}{2}\,\ell n\, a + \tfrac{1}{4} - \frac{I_0(\sigma)}{2\sigma I_1(\sigma)}\}, \tag{26.25}$$

where σ is given by (24.7). The outer limit of the inner asymptotic expansion is found from (26.22)

$$S(r) \simeq \beta - \tfrac{1}{2} L\,a^2\,\ell n\, r \qquad (a \ll r \ll h\eta/\eta'). \tag{26.26}$$

The unknown constant L can now be determined as follows. In the regime $a \ll r \ll h\eta/\eta'$ both asymptotic solutions (26.18) and (26.26) are valid. This gives two equations between the two unknown quantities L and F, which can be solved to give

$$F = 4\pi\eta h v_0\,\{-\gamma + \ell n\,(\frac{\eta h}{\eta' a}) + \frac{2}{\sigma^2} + \frac{I_0(\sigma)}{\sigma I_1(\sigma)}\}^{-1}, \tag{26.27}$$

$$La^2 = 2v_0\,\{-\gamma + \ell n\,(\frac{\eta h}{\eta' a}) + \frac{2}{\sigma^2} + \frac{I_0(\sigma)}{\sigma I_1(\sigma)}\}^{-1}. \tag{26.28}$$

These results are correct if $\eta h/\eta' a \gg 1$, hence in this regime the translational friction coefficient is given by

$$f_T \simeq 4\pi\eta h\,\{-\gamma + \ell n\,(\frac{\eta h}{\eta' a}) + \frac{2}{\sigma^2} + \frac{I_0(\sigma)}{\sigma I_1(\sigma)}\}^{-1}. \tag{26.29}$$

The Einstein relation (8.1) gives for the translational diffusion coefficient

$$D_T \simeq \frac{k_B T}{4\pi\eta h}\,\{-\gamma + \ell n\,(\frac{\eta h}{\eta' a}) + \frac{2}{\sigma^2} + \frac{I_0(\sigma)}{\sigma I_1(\sigma)}\}, \tag{26.30}$$

which result too holds asymptotically for $h\eta/a\eta' \gg 1$. These formulae are from Wiegel (1979a,b,c).

In the limit $\sigma \to \infty$ the cylindrical disk becomes impermeable and we recover the result of Saffman and Delbruck (1975) and Saffman (1976) for the translational friction coefficient of a hard disk

$$f_T(\infty) \simeq 4\pi\eta h\,\{-\gamma + \ell n\,(\frac{\eta h}{\eta' a})\}^{-1}. \tag{26.31}$$

In the free draining regime one finds

$$f_T \cong \pi\eta h\sigma^2, \qquad (\sigma \ll 1). \qquad (26.32)$$

The correction term $\frac{2}{\sigma^2} + \frac{I_0(\sigma)}{\sigma I_1(\sigma)}$ due to the finite permeability of the disk is tabulated in table VI.

σ	$\frac{2}{\sigma^2} + \frac{I_0(\sigma)}{\sigma I_1(\sigma)}$
0	∞
1	4.2402
2	1.2166
3	0.6337
4	0.4145
5	0.3039
6	0.2382
7	0.1952
8	0.1649
9	0.1426
10	0.1254
∞	0

Table VI. Correction term in the translational transport coefficients (26.29,30) due to a finite permeability of a disk.

27. EXPERIMENTS ON MEMBRANE DIFFUSION

As expounded in part C, the experimental determination of the diffusion coefficients of globular proteins and permeable patches in the cell membrane is of considerable interest.

We now briefly review the data. When comparing with the theoretical predictions we always shall use the values of η, η' and h given by (19.1-3).

The smallest objects found in the membrane are the lipids themselves. Although the lipid bilayer is described in our model as a continuum we can, nevertheless, extrapolate (26.30) to hold for a single lipid. Using

$$k_BT \simeq 4.14 \times 10^{-14}\ \mathrm{cm}^2\ \mathrm{g}\ \mathrm{s}^{-2} \tag{27.1}$$

at $T = 300\ ^0K \simeq 27\ ^0C$, and (19.4) for the radius of a lipid, eq. (26.30) with $\sigma = \infty$ gives

$$D_T \simeq 2.9 \times 10^{-8}\ \mathrm{cm}^2\ \mathrm{s}^{-1}. \tag{27.2}$$

The translational diffusion coefficient of lipids in synthetic black lipid membranes has been measured by Fahey et al. (1977); these authors find

$$D_T \simeq 2 \times 10^{-7}\ \mathrm{cm}^2\ \mathrm{s}^{-1} \tag{27.3}$$

at 25 ^0C. The experiment has been analyzed recently by Webb (1978) who argues that the unusually high diffusion coefficient of lipids in black lipid membranes is due to retained solvent. In this same paper Webb gives for the diffusion coefficient of lipids in natural cell membranes values of the order of magnitude

$$D_T \simeq 1 \times 10^{-8}\ \mathrm{cm}^2\ \mathrm{s}^{-1}, \tag{27.4}$$

in fair agreement with the theoretical result (27.2). The experimental data (27.4) are taken from Schlessinger et al. (1976), Schlessinger et al. (1977), and from Fahey and Webb (1978).

For the record, the leading factor in (26.30) and (24.10) has the value

$$\frac{k_BT}{4\pi\eta h} \simeq 0.41 \times 10^{-8}\ \mathrm{cm}^2\ \mathrm{s}^{-1} \tag{27.5}$$

at 300 ^0K. A glance at these two equations shows that D_R strongly depends on the radius, roughly like a^{-2}, but that D_T depends on the radius only through a logarithmic term.

Both the rotational- and the translational diffusion coefficient have been measured for rhodopsin in the frog photoreceptor membrane (Cone, 1972; Poo and Cone,

1974)

$$D_T \cong 3.5 \times 10^{-9} \text{ cm}^2 \text{ s}^{-1}, \tag{27.6}$$

$$D_R \cong 0.5 \times 10^5 \text{ s}^{-1} \tag{27.7}$$

at 20 ^{0}C. This protein has a radius

$$a \cong 20 \text{ Å}. \tag{27.8}$$

The theoretical estimations are

$$D_T \cong 2.2 \times 10^{-8} \text{ cm}^2 \text{ s}^{-1}, \tag{27.9}$$

$$D_R \cong 1 \times 10^5 \text{ s}^{-1}. \tag{27.10}$$

In view of the considerable uncertainty in the experimental values, agreement between theory and experiment is fair. The fact that the theoretical values are somewhat too large could be due to either of the following two effects: 1. Some lipids are tightly bound to the rhodopsin molecule. This would lead to a larger value of a, hence to smaller values of D_T and D_R. 2. The rhodopsin molecule is tethered, either to the cytoskeleton or to other membrane proteins.

Even smaller diffusion coefficients have been measured for integral proteins in the human erythrocyte membrane. Fowler and Branton (1977) find

$$D_T \cong 3 \times 10^{-11} \text{ cm}^2 \text{ s}^{-1} \tag{27.11}$$

at 30 ^{0}C, and Cherry et al. (1976) find

$$D_R \cong 1 \times 10^3 \text{ s}^{-1} \tag{27.12}$$

at 22 ^{0}C. These proteins have a radius which is estimated to be approximately 40 Å. The measured diffusion coefficients are now considerably smaller than the predicted values. This almost certainly implies that these proteins are attached to other objects inside the cell. Some of the effects of tethering of proteins through cross-linking with antigens have been studied experimentally by Wolf et al. (1977), but to date no theoretical studies have been completed.

28. DISCUSSION AND OUTLOOK

It seems appropriate to conclude this booklet with a list of problems which could be fruitful topics for further research. The general physical situation of fluid flow in a porous medium is of considerable practical importance and much useful work can still be done.

For the historian of science there is the remarkable fact that the publication by Darcy (1856) in which the hydrodynamic permeability was first introduced appeared almost simultaneous with an early paper by Maxwell (1856) on the electromagnetic field, in which he developed an analogy for lines of force in terms of fluid flow through a resistive medium (compare the discussion of Maxwell's ideas by Wise, 1979). One wonders if the hydrodynamic permeability served as a model for its electromagnetic counterparts?

At the end of section 7 we listed the three assumptions under which the fundamental equation (5.5) has been derived. If these assumptions are not satisfied most of our results could break down. For example, in situations in which the relevant Reynolds numbers are not small as compared to unity the non-linear term in the Navier-Stokes equation has to be retained and the linear structure of the theory is lost. Although this does not happen in typical biophysical applications (compare the estimates in section 1) other applications are possible in which the Reynolds numbers are not small.

When the fraction of space occupied by the medium is not negligeably small the Felderhof-Deutch formalism of section 7 should be corrected with terms which result from the boundary conditions at the fluid-medium interface. This question is related to the problem of calculating the concentration dependence of f_R, f_T and $[\eta]$. This problem is very difficult but progress has been made recently by Felderhof (1976a,b), Jones (1978a,b,c; 1979) and Reuland, Felderhof and Jones (1978).

The mean-field approximation (7.19) is ad-hoc. Still, in view of the macroscopic derivation of the fundamental equation given in section 5 one expects that this is an excellent approximation as long as the problem is dominated by two lengths ℓ and L such that the order of magnitude estimate (1.6) holds. However, when the macroscopic scale L becomes comparable to the microscopic scale ℓ our coarse-grained description will break down. This will only be the case in a shock-wave or under similar extreme circumstances.

In these lectures the permeable medium has always been represented as a rigid object, similar to a piece of chalk. Under certain conditions the medium will suffer deformations of its own. In those cases one could describe the medium as being elastic and governed by the equations of elasticity theory. This description would lead to a set of coupled partial differential equations for two time-dependent velocity fields $\vec{U}(\vec{r},t)$ and $\vec{V}(\vec{r},t)$

$$\rho_0 \frac{\partial \vec{V}}{\partial t} = -\vec{\nabla}\, P + \eta_0 \Delta \vec{V} + \frac{\eta_0}{k}(\vec{U} - \vec{V}), \qquad (28.1)$$

$$\text{div}\, \vec{V} = 0 \qquad (28.2)$$

$$\rho_1 \frac{\partial \vec{U}}{\partial t} = \vec{F}[\vec{U},\vec{V}], \qquad (28.3)$$

where $\vec{F}$ denotes the total force exerted on the elastic medium, per unit volume. The explicit form of the functional $\vec{F}[\vec{U},\vec{V}]$ depends on the rheological details of the medium; in any case $\vec{F}$ will contain an additive term $\frac{\eta_0}{k}(\vec{V} - \vec{U})$ due to the interaction with the fluid. To the author's knowledge no applications of this type have been made.

To date no one has calculated f_R and D_R of Gaussian coils. A numerical solution of this problem should be straightforward. In general it should not be too difficult to write a computer program for polymer coils in a fluid with input k(r), the radial dependence of the hydrodynamic permeability, and with output D_R, D_T and $[\eta]$.

It also could be of interest to contemplate the inverse problem in which one gives the values of, for example, $D_T(m)$ for polymers of the same type but different molecular mass m and asks for the function k(r). This might be a "cheap" way to determine the molecular characteristics of polymer coils.

It would be of some interest to generalize the calculation of the rotational friction coefficient of a uniform disk in sections 23 and 24 to the case of non-zero values of $a\eta'/h\eta$. This leads to a set of dual integral equations which are not easy to solve. A more interesting question is to generalize the calculation in section 26 of $D_T(a\eta'/h\eta)$ to all values of its argument; from a biophysical point of view the regime in which $a\eta'/h\eta$ is of order unity is of interest. This problem also leads to dual integral equations.

Finally one should extend the calculations in part C in such a way that the effects of the "tethering" of proteins are taken into account. This problem is related to the question of whether a system of filaments is present inside the cell which can actually exert forces on specific objects in the cell membrane. These forces might play an essential role in cap formation and other events in the immune system.

APPENDIX. COMMENTS ON ENTANGLEMENTS AND THE EXCLUDED VOLUME PROBLEM

A. Importance of these problems. The aim of this appendix is to supplement the material in the book with comments pertinent to certain characteristic problems which arise in the statistical mechanics of macromolecules. These problems, which are caused by the linear character of a macromolecule, are related to questions of a topological nature and have not (yet) been encountered in other branches of theoretical physics.

The simplest problem in this category consists of calculating the probability that a macromolecule will be n-times entangled with a straight line. This "simple entanglement problem" arises from the mutual entanglement of two chains when one chain is replaced by a straight line. The simple entanglement problem, and some of its generalizations, has been studied in detail by Prager and Frisch (1967), Edwards (1967, 1978), Saito and Chen (1973), Wiegel (1977b) and Alexander-Katz and Edwards (1972); a review can by found in section V of Wiegel (1979d).

A second problem which is characteristic for the statistical mechanics of macromolecules is the excluded volume problem. This problem, which consists of calculating the corrections to the free random walk chain statistics (13.8,10) due to the finite volume of a macromolecule, has already been discussed briefly at the end of section 13 where we cited the review papers of McKenzie (1976) and of Lifshitz, Grosberg and Kokhlov (1978).

Recently it has been shown that a profound relation exists between certain entanglement problems and the two-dimensional excluded volume problem (Wiegel 1979e,f). In order to put this relation on a firm basis one has to consider the entanglement problem for two-dimensional random walks with a complex weight. It is remarkable that two-dimensional random walks with exactly the same complex weights play a role in the combinatorial solution of the Ising model and the "free-fermion" model, as has been shown by Kac and Ward (1952), Sherman (1960, 1963), Burgoyne (1963), Vdovichenko (1965) and Wiegel (1972, 1975a).

B. The two-dimensional simple entanglement problem. The macromolecule is represented by N freely hinged rods, each of fixed length ℓ, with end points at positions $\vec{r}_0, \vec{r}_1, \ldots \vec{r}_N$. We consider only the two-dimensional case in which the molecule is constrained to a plane perpendicular to the line and in which we calculate the configuration sum $Q_n(\vec{r}_N, N|\vec{r}_0, 0)$ over those configurations which are n-times entangled with the origin

$$Q_n(\vec{r}_N,N|\vec{r}_0,0) = \int d^2\vec{r}_1 \int d^2\vec{r}_2 \ldots \int d^2\vec{r}_{N-1}\, E_n\,(\vec{r}_0,\vec{r}_1,\ldots\vec{r}_N) \prod_{i=0}^{N-1} (2\pi\ell)^{-1}\, \delta(|\vec{r}_{i+1}-\vec{r}_i|-\ell), \tag{B.1}$$

where $E_n(C) = 1$ if the configuration $C \equiv (\vec{r}_0, \vec{r}_1, \ldots \vec{r}_N)$ is n-times entangled with the origin, and $E_n(C) = 0$ in all other cases. In order to precisely define the number of times C is entangled with the origin we draw some continuous curve T which starts at the origin, does not contain any self-intersections and ends at infinity. Let $n_+(C)$ denote the number of times that C crosses T going in one direction (the "positive" direction) and let $n_-(C)$ denote the number of times that C crosses T going in the opposite "negative" direction. The entanglement index n is defined as

$$n(C) = n_+(C) - n_-(C), \tag{B.2}$$

and the molecule is said to be n-times entangled with the origin.

In this case, if the configuration $(\vec{r}_0, \vec{r}_1, \ldots, \vec{r}_{N-1}, \vec{r}_N)$ has an entanglement index n, then the configuration $(\vec{r}_0, \vec{r}_1, \ldots \vec{r}_{N-1})$ has the same entanglement index n, unless $\vec{r}_N$ has a distance to T which is smaller than ℓ. This leads to the integral equation

$$Q_n(\vec{r}_N, N|\vec{r}_0, 0) = (2\pi\ell)^{-1} \int Q_n(\vec{r}_{N-1}, N-1|\vec{r}_0, 0)\ \delta|\vec{r}_N - \vec{r}_{N-1}| - \ell)\, d^2\vec{r}_{N-1}. \tag{B.3}$$

For $N \gg 1$ the dependence of the functions $Q_n(\vec{r}_N, N|\vec{r}_0, 0)$ on N and $\vec{r}_N$ will be smooth, hence each of these functions can be expanded in a Taylor series around the point $(\vec{r}_N, N)$. This procedure leads to the differential equation

$$\left[\frac{\partial}{\partial N} - \frac{\ell^2}{4} \Delta_{\vec{r}_N}\right] Q_n(\vec{r}_N, N|\vec{r}_0, 0) = 0, \tag{B.4}$$

where $\Delta_{\vec{r}_N} \equiv \partial^2/\partial x_N^2 + \partial^2/\partial y_N^2$. The initial condition is

$$\lim_{N \downarrow 0} Q_n(\vec{r}_N, N|\vec{r}_0, 0) = \begin{cases} \delta(\vec{r}_N - \vec{r}_0), & (n=0), \\ 0 & (n \neq 0). \end{cases} \tag{B.5}$$

It is important to note that the functions Q_n are related to each other by boundary conditions which hold at the curve T. These conditions can be written in the form

$$\lim_{\vec{r}_N \uparrow T} Q_n(\vec{r}_N, N|\vec{r}_0, 0) = \lim_{\vec{r}_N \downarrow T} Q_{n+1}(\vec{r}_N, N|\vec{r}_0, 0), \tag{B.6}$$

where the symbol $\vec{r}_N \uparrow T$ indicates the limit in which $\vec{r}_N$ approaches some point $\vec{r} \in T$ in such a way that $\vec{r}_N$ moves in the positive direction; similarly $\vec{r}_N \downarrow T$ means that $\vec{r}_N$ approaches the same point $\vec{r} \in T$ in such a way that $\vec{r}_N$ now moves in the negative direction. Relations similar to (B.6) also hold for all derivatives.

In order to calculate the functions Q_n we introduce polar coordinates: $\vec{r}_N \equiv (r, \theta)$ with $0 < r < \infty$, $-\pi + 2\pi n < \theta \leq +\pi + 2\pi n$. The curve T is choosen to correspond to $\theta = +\pi + 2\pi n$; $0 < r < \infty$. Writing $Q_n(r, \theta, N)$ for the function $Q_n(\vec{r}_N, N|\vec{r}_0, 0)$ we can introduce a function $q(r, \theta, N)$ which is defined for $-\infty < \theta < +\infty$ by

$$q(r,\theta,N) = Q_n(r,\theta,N), \quad -\pi + 2\pi n < \theta \leq +\pi + 2\pi n, \quad (n = 0,\pm 1,\pm 2,\ldots). \tag{B.7}$$

The boundary conditions (B.6) and the equations (B.4) guarantee that $q(r,\theta,N)$ will satisfy the differential equation

$$\left[\frac{\partial}{\partial N} - \frac{\ell^2}{4}\left(\frac{\partial^2}{\partial r^2} + \frac{1}{r}\frac{\partial}{\partial r} + \frac{1}{r^2}\frac{\partial^2}{\partial \theta^2}\right)\right] q(r,\theta,N) = 0 \tag{B.8}$$

for $-\infty < \theta < +\infty$. Note that another choice for T would have led to the same equation (B.8) provided the regime of validity $-\pi + 2\pi n < \theta \leq +\pi + 2\pi n$ of (B.7) is modified in the appropriate way. The boundary conditions on q are

$$q(0,\theta,N) = 0, \quad (-\infty < \theta < +\infty), \tag{B.9}$$

$$q(\infty,\theta,N) = 0, \quad (-\infty < \theta < +\infty), \tag{B.10}$$

$$q(r,\pm\infty,N) = 0, \quad (0 < r < +\infty). \tag{B.11}$$

It is now straightforward to show (see Wiegel 1977b for details) that q can be expanded in the eigenfunctions of the differential operator on the left hand side of (B.8). One finds for the configuration sum over the n-times entangled configurations

$$Q_n(r,\theta,N) = \frac{1}{2\pi}\int_{-\infty}^{+\infty} dk \sum_m F_m(r)F_m^*(r_0)\exp\{-\lambda_m(k)N + (\theta-\theta_0)ki\}. \tag{B.12}$$

In this equation $\vec{r}_0$ has polar coordinates (r_0,θ_0) with $-\pi < \theta_0 \leq +\pi$ and θ lies in the interval $-\pi + 2\pi n < \theta \leq +\pi + 2\pi n$. The $F_m(r)$ have to be solved from the equations

$$\left[\frac{d^2}{dr^2} + \frac{1}{r}\frac{d}{dr} - \frac{k^2}{r^2} + \frac{4}{\ell^2}\lambda_m\right] F_m(r) = 0, \tag{B.13}$$

$$F_m(0) = 0, \qquad F_m(\infty) = 0, \tag{B.14,15}$$

$$\int_0^\infty r\, F_m(r)F_{m'}^*(r)dr = \delta_{m,m'}. \tag{B.16}$$

The solution of (B.13,14) can be given in terms of Bessel functions of the first kind

$$F_m(r) = A\, J_{|k|}\left(\frac{r}{\ell}\sqrt{4\lambda_m}\right), \tag{B.17}$$

where A is a constant. In order to determine λ_m we impose the boundary condition $F_m(R) = 0$ at some very large radius R; at the end of the calculation $R \to \infty$. If $x_{p,m}$ denotes the mth zero on the positive real x axis of $J_p(x) = 0$ this boundary condition gives

$$\frac{R}{\ell}\sqrt{4\lambda_m} = x_{|k|,m} , \tag{B.18}$$

so the eigenvalues are

$$\lambda_m(k) = \frac{\ell^2}{4R^2} x^2_{|k|,m} . \tag{B.19}$$

In order to evaluate the summation over m in (B.12) one has to determine the normalization constant A, which follows from

$$\int_0^R r J_{|k|} (\tfrac{r}{R} x_{|k|,m}) J_{|k|} (\tfrac{r}{R} x_{|k|,m'}) \, dr = A^{-2} \delta_{m,m'} . \tag{B.20}$$

The integral is given in eq. 6.521.1 of Gradshteyn and Ryzhik (1965): the integral vanishes for $m \neq m'$; for $m = m'$ one has

$$\int_0^R r J^2_{|k|}(\tfrac{r}{R} x_{|k|,m}) \, dr = \tfrac{1}{2} R^2 J^2_{|k|+1}(x_{|k|,m}) . \tag{B.21}$$

The normalized eigenfunctions are now given by

$$F_m(r) = \frac{\sqrt{2}}{R} \frac{J_{|k|} (\frac{r}{R} x_{|k|,m})}{|J_{|k|+1} (x_{|k|,m})|} . \tag{B.22}$$

Upon substitution of this expression and (B.19) into (B.12) one is led to evaluate the series

$$\sum_m F_m(r) F^*_m(r_0) \exp(-\lambda_m N) = \frac{2}{R^2} \sum_{m=1}^{\infty} \frac{J_{|k|} (\frac{r}{R} x_{|k|,m}) \, J_{|k|} (\frac{r_0}{R} x_{|k|,m})}{J^2_{|k|+1} (x_{|k|,m})} \cdot$$

$$\cdot \exp(- \frac{\ell^2 N}{4R^2} x^2_{|k|,m}) . \tag{B.23}$$

At this point it is convenient to take the limit $R \to \infty$. In this limit the sum over m will be dominated by those terms for which $x_{|k|,m}$ is of order $R/\ell\sqrt{N} >> 1$. This enables us to use the asymptotic formula

$$J_\nu(x) \simeq \sqrt{\frac{2}{\pi x}} \cos(x - \tfrac{1}{2}\pi\nu - \tfrac{\pi}{4}), \qquad (x >> 1). \tag{B.24}$$

The zero's are asymptotically given by

$$x_{\nu,m} \simeq \tfrac{1}{2}\pi\nu + \tfrac{\pi}{4} + (m-1)\pi + \tfrac{\pi}{2} , \tag{B.25}$$

hence

$$J^2_{|k|+1}(x_{|k|,m}) \cong \frac{2}{\pi x_{|k|,m}} . \tag{B.26}$$

For large values of R the variable

$$\zeta = \frac{\ell}{R} x_{|k|,m} \tag{B.27}$$

behaves as a continuous variable. If $g(\zeta)d\zeta$ denotes the number of ζ-values in the interval $d\zeta$ one finds

$$g(\zeta) \cong \frac{R}{\pi\ell} , \qquad (R/\ell \gg 1). \tag{B.28}$$

Substituting these results into (B.23) we find

$$\lim_{R\to\infty} \sum_m F_m(r)\, F^*_m(r_0) \exp(-\lambda_m N) =$$

$$= \ell^{-2} \int_0^\infty \zeta\, J_{|k|}(\frac{r}{\ell}\zeta)\, J_{|k|}(\frac{r_0}{\ell}\zeta) \exp(-\tfrac{1}{4}N\zeta^2)\, d\zeta. \tag{B.29}$$

This integral can be found as eq. 6.633.2 of Gradshteyn and Ryzhik (1965) and the right hand side of (B.29) equals

$$\frac{2}{N\ell^2} \exp(-\frac{r^2+r_0^2}{N\ell^2})\, I_{|k|}(\frac{2r_0 r}{N\ell^2}), \tag{B.30}$$

where the I_ν are the modified Bessel functions. Combinations of (B.12) and (B.29,30) gives the solution of the two-dimensional simple entanglement problem in the form of a definite integral

$$Q_n(r,\theta,N) = (\pi N\ell^2)^{-1} \exp(-\frac{r_0^2+r^2}{N\ell^2}) .$$

$$\int_{-\infty}^{\infty} I_{|k|}(\frac{2r_0 r}{N\ell^2}) \exp\{(\theta-\theta_0)k\, i\}\, dk. \tag{B.31}$$

Some applications of this result can be found in a forthcoming paper by the author (Wiegel, 1979d). We shall now consider the excluded volume problem, returning to the solution (B.31) in a later stage of our discussion.

C. Role of dimension in the excluded volume problem. At the end of section 13 we invoked a qualitative argument which indicated the role of the dimensionality d. For $d > 4$ the excluded volume effect is negligible if $N \gg 1$; for $d < 4$ the self-interaction of a polymer dominates the statistics of its configurations. The case

$d = 4$ is somewhat marginal in the sense that if some effect is present it should lead to a quantitative rather than a qualitative change in the chain statistics. As the one-dimensional problem is trivial the cases $d = 2$ and $d = 3$ represent the only cases in which the excluded volume problem is a relevant problem.

D. The critical exponent ν. For historical reasons the theory has concentrated on the calculation of the "critical exponent" ν which is defined by

$$<|\vec{R}|^2> \cong A\ell^2 N^{2\nu} , \qquad (N >> 1), \qquad (D.1)$$

where R denotes the vector connecting the two end points of the chain and where the average is taken over all the self-avoiding chain configurations; A is a numerical constant. The self-avoiding walk constraint models the purely geometric steric hindrance of the polymer chain, i.e. the chain is represented by a "necklace" of hard spheres. For ideal chains $\nu = \frac{1}{2}$ in any dimension. Numerical experiments (McKenzie,1976) suggest that $\nu \cong 0.60$ for chains with excluded volume in three dimensions. We shall now consider a self-consistent field approach to find an approximate solution to the excluded volume problem.

E. The self-consistent field approximation. Consider all self-avoiding random walks of N steps, in a d-dimensional continuous space, which start in the origin of space. Around each endpoint we imagine a small hard sphere of radius $\frac{a}{2} < \frac{\ell}{2}$, i.e. around each endpoint there is a volume γ_d which is forbidden for all other endpoints. Here $\gamma_2 = \pi a^2$, $\gamma_3 = \frac{4}{3}\pi a^3$, etc. Let $f(\vec{r})$ denote the average volume fraction which is thus excluded; the average is taken over all these self-avoiding walks. This function will have radial symmetry around the origin: $f(\vec{r}) = f(r)$, and will be monotonically decreasing at large distances.

The self-consistent field approach assumes that the statistical properties of the self-avoiding walks are the same as those of free random walks in a medium in which a volume fraction $f(r)$ of space is occupied by some perfectly absorbing material. The statistical properties of this second system can be calculated as follows.

The probability that a specific free random walk $\vec{0}, \vec{r}_1, \vec{r}_2, \ldots \vec{r}_N$ does not hit upon absorbing material anywhere along its path equals

$$\prod_{i=1}^{N} (1-f(\vec{r}_i)) \cong \exp \{- \int_0^N f(\vec{r}(n))dn\}, \qquad (E.1)$$

where we used the fact that $f<<1$ and the fact that f will be practically constant over distances of order ℓ provided $N>>1$. The number of free random walk configurations in which the shape of the polymer is near to the curve $\vec{r}(n)$ is proportional to

$$\exp \{- \frac{1}{4D_d} \int_0^N (\frac{d\vec{r}}{dn})^2 \, dn\},\tag{E.2}$$

where $D_2 = \frac{1}{4}\ell^2$, $D_3 = \frac{1}{6}\ell^2$ (compare Wiegel, 1975b, section 3). Hence, the number of free random walks of shape $\vec{r}(n)$ which do not hit any absorbing material is proportional to

$$\exp \{- \int_0^N [\frac{1}{4D_d} (\frac{d\vec{r}}{dn})^2 + f(\vec{r})] \, dn\}.\tag{E.3}$$

The most probable configuration ($\vec{r}^*(n)$) of the polymer in this absorbing medium is found by minimizing the exponential; it is a solution of the diffential equation

$$\frac{1}{2D_d} \frac{d^2\vec{r}^*}{dn^2} = \frac{\partial f(\vec{r}^*)}{\partial \vec{r}^*},\tag{E.4}$$

under the boundary condition

$$\vec{r}^*(0) = \vec{0}.\tag{E.5}$$

This is the equation of motion of a fictitious classical particle with coordinate $\vec{r}^*$, mass $(2D_d)^{-1}$ and time n, moving in an external potential equal to $-f(\vec{r}^*)$. The total energy E is a constant of the motion, hence

$$\frac{1}{4D_d} (\frac{d\vec{r}^*}{dn})^2 = f(\vec{r}^*) + E.\tag{E.6}$$

The motion of the particle proceeds along a radius vector in the direction of $(d\vec{r}^*/dn)_{n=0}$. The value of E has to be determined from the constraint on the total length of the chain

$$N = \int_0^{r_{max}} \frac{dr}{\sqrt{4D_d(f+E)}},\tag{E.7}$$

where r_{max} follows from

$$f(r_{max}) = -E.\tag{E.8}$$

Hence E will be negative and very small for $N \gg 1$. Putting $E \cong 0$ in (E.6) one finds

$$\frac{dr^*}{dn} = \sqrt{4D_d f(r^*)}.\tag{E.9}$$

The function f(r) can now be determined in a self-consistent way by the following argument attributable to de Gennes (1969). Consider a spherical shell around the origin with radius r and thickness dr. The average number of monomers inside this shell equals $(A_d/\gamma_d)r^{d-1}f(r)dr$ where $A_d\, r^{d-1}$ denotes the area of a d-dimensional sphere of radius r. On the other hand, the average number of monomers is approximately equal to the number of monomers of the most probable configuration $r^*(n)$ which are inside this shell; this number equals $dn = dr(dr^*/dn)^{-1}$. One finds

$$\frac{A_d}{\gamma_d}\, r^{d-1} f(r) = \{4D_d f(r)\}^{-\frac{1}{2}} \tag{E.10}$$

so that

$$f(r) = \left(\frac{\gamma_d}{A_d\sqrt{4D_d}}\right)^{\frac{2}{3}} r^{\frac{2}{3}(1-d)} . \tag{E.11}$$

The most probable trajectory is now found upon substitution into (E.9)

$$\frac{dr^*}{dn} = \sqrt{4D_d}\left(\frac{\gamma_d}{A_d\sqrt{4D_d}}\right)^{\frac{1}{3}} r^{*\frac{1}{3}(1-d)} . \tag{E.12}$$

Solving this differential equation for $r^*(n)$ one finds

$$r^{*\frac{1}{3}(d+2)} \sim n, \tag{E.13}$$

where the proportionality constant has not been written down explicity. Thus, the most probable trajectories have the form

$$r^*(n) \sim n^{3/(d+2)} . \tag{E.14}$$

Putting $n = N$, comparing with (D.1) and always identifying the average over the chain configurations with the behavior of the most probable chain configuration one finds the value

$$\nu = \frac{3}{d+2} \qquad (d \leq 4). \tag{E.15}$$

For $d > 4$ this would give a value smaller than the free random walk value $\nu = \frac{1}{2}$, hence (E.15) has to be amended with

$$\nu = \tfrac{1}{2} \qquad (d \geq 4). \tag{E.16}$$

The ν-values 1, $\frac{3}{4}$ and $\frac{3}{5}$ for 1, 2 and 3 dimensions are in satisfactory agreement with computer enumerations (compare McKenzie (1976) and the papers quoted in that reference). An older derivation of (E.15) is due to Edwards (1965a,b) and uses the somewhat esoteric language of path integration; for this approach the reader might find it instructive to compare Wiegel (1975b). All these calculations rely on two ad-hoc assumptions: (1) The constraint of self-avoidingness is replaced by an absorbing medium; (2) The average of a function over all chain configurations is replaced by the value of the function for the most probable configuration.

F. The cluster expansion. More than ten years ago the author attempted to get a clue to the solution of the excluded volume problem by studying the cluster expansion for a macromolecule with self-interaction (Wiegel, 1969). The idea was to expand the configuration sum

$$G(\vec{r}_N,N|\vec{r}_0,0) = \int d^3\vec{r}_1 \int d^3\vec{r}_2 \ldots \int d^3\vec{r}_{N-1} \exp\{-\beta \sum_{i<j} V(\vec{r}_i-\vec{r}_j)\} \cdot \prod_{i=0}^{N-1} (4\pi\ell^2)^{-1} \delta(|\vec{r}_{i+1}-\vec{r}_i|-\ell) \tag{F.1}$$

in a series in terms of the Mayer function

$$f(\vec{r}) \equiv \exp\{-\beta V(\vec{r})\} -1. \tag{F.2}$$

Here $\beta = (k_B T)^{-1}$ and $V(\vec{r})$ denotes the potential of the interaction between the end points of the N links. The resulting cluster diagrams have been discussed elsewhere (Wiegel 1975b, section 3).

The cluster series leads to a "Dyson equation" in the following way. Take the Fourier transform

$$\tilde{G}(k,N) \equiv \int G(\vec{r},N|\vec{0},0) \exp(i\vec{k}\cdot\vec{r}) d^3\vec{k} \tag{F.3}$$

and the generating function

$$\tilde{G}(k,z) \equiv \sum_{N=1}^{\infty} \tilde{G}(k,N) z^N. \tag{F.4}$$

These functions depend only on $k \equiv |\vec{k}|$ because of the spherical symmetry of $G(\vec{r},N|\vec{0},0)$. It can now be shown that a Dyson equation holds which has the form

$$\tilde{G}(k,z) = \tilde{G}_0(k,z) \{1-\tilde{\Pi}(k,z)\tilde{G}_0(k,z)\}^{-1}. \tag{F.5}$$

Here $\tilde{G}_0$ denotes the equivalent of $\tilde{G}$ in the free random walk model and

$$\tilde{\Pi}(k,z) \equiv \int \Pi(\vec{r},z) \exp(i\vec{k}\cdot\vec{r}) d^3\vec{r} = (4\pi/k) \int_0^{\infty} r\Pi(r,z) \sin kr \, dr. \tag{F.6}$$

The function $\Pi(\vec{r},z)$ is defined as a series of irreducible cluster diagrams, i.e. $\Pi(\vec{r},z)$ equals the sum of all cluster diagrams which cannot be separated into two disconnected diagrams by cutting a single propagator line.

G. Conjecture about the non-Gaussian statistics. It can be seen from (F.3) that for $r/\ell >> 1$ the contributions to the configuration sum $G(\vec{r},N|\vec{0},0)$ come from those values of $\vec{k}$ for which $k\ell << 1$. We can replace $\tilde{\Pi}(k,z)$ by the first two terms in its Taylor series (assuming that such an expansion exists)

$$\tilde{\Pi}(k,z) = \tilde{\Pi}(0,z) + \tfrac{1}{2} k^2 \tilde{\Pi}''(0,z), \tag{G.1}$$

where

$$\tilde{\Pi}(0,z) = 4\pi \int_0^\infty r^2 \Pi(r,z) dz \tag{G.2}$$

and

$$\tilde{\Pi}''(0,z) = -4\pi \int_0^\infty r^4 \Pi(r,z)\, dz \tag{G.3}$$

are unknown functions of z. Note that (13.7) implies

$$\tilde{G}_0^{-1}(k,z) = \frac{1}{z} \exp\left(\tfrac{1}{6}k^2\ell^2\right) - 1. \tag{G.4}$$

Substitution of (G.4) and (G.1) into (F.5) gives

$$\tilde{G}^{-1}(k,z) \cong \left\{\frac{1}{z} - 1 - \tilde{\Pi}(0,z)\right\} + \left\{\frac{\ell^2}{6z} - \tfrac{1}{2}\tilde{\Pi}''(0,z)\right\}k^2, \tag{G.5}$$

provided $k\ell << 1$. The configuration sum is given by the inverse Fourier transform

$$G(r,z) = (2\pi)^{-3} \int \frac{e^{-i\vec{k}\vec{r}}}{\left\{\frac{1}{z} - 1 - \tilde{\Pi}(0,z)\right\} + \left\{\frac{\ell^2}{6z} - \frac{1}{2}\tilde{\Pi}''(0,z)\right\}k^2}\, d^3\vec{k}$$

$$= (4\pi r\gamma(z))^{-1} \exp\{-r\phi(z)\}, \tag{G.6}$$

where

$$\phi(z) = \left\{ \frac{1 - z - z\tilde{\Pi}(0,z)}{\ell^2/6 - \frac{1}{2} z \tilde{\Pi}''(0,z)} \right\}^{\frac{1}{2}}, \tag{G.7}$$

$$\gamma(z) = \frac{\ell^2}{6z} - \frac{1}{2}\tilde{\Pi}''(0,z). \tag{G.8}$$

It is now tempting to conjecture where the non-Gaussian statistics come from (Wiegel, 1969). Let us assume that the singularity of the function $\phi(z)$ which is nearest to the origin of the complex z-plane consists of a branchline along the positive real z-axis from z_0 to ∞. Suppose ϕ behaves asymptotically as

$$\phi(z) \simeq a(z_0-z)^b, \qquad (z \uparrow z_0). \tag{G.9}$$

In this case G(r,N) can be calculated using the saddlepoint method

$$G(r,N) = (2\pi i)^{-1} \oint G(r,z)z^{-N-1}dz. \tag{G.10}$$

One finds that the leading terms are given by

$$G(r,N) \simeq c^N \exp\{-d\,(\frac{r}{\ell N^b})^{\frac{1}{1-b}}\}, \qquad (N, \frac{r}{\ell} >> 1), \tag{G.11}$$

where c and d are constants. Hence the non-Gaussian statistics can be represented by a function of the "scaled" distance $\frac{r}{\ell N^b}$, and $b \simeq 0.60$ would lead to agreement with the experimental estimates.

H. The renormalization group approach. The scaling properties of a polymer with excluded volume have been studied extensively, during the last five years, with the method of the renormalization group. This method, which leads to rather accurate results for the critical exponent ν and related quantities, lies outside the scope of this appendix. Some of the more important older contributions are found in de Gennes (1972), des Cloizeaux (1974,1975), Emery (1975) and Hilhorst (1976). This part of the literature is still expanding rapidly.

I. Markovian character of the two-dimensional excluded volume problem. It has recently been shown by the author (Wiegel, 1979e,f) that a profound relation exists between certain entanglement problems and the two-dimensional self-avoiding random walk problem. This relation is a byproduct of a demonstration of the Markovian nature of the latter problem. The Markovian nature of the two-dimensional self-avoiding random walk problem can be shown by a combination of two methods which have been used in the past to count random walks subject to certain global constraints:
(a) the combinatorial method, which has been used to solve a certain class of models for cooperative behavior in two dimensions; (b) a method used to count random walks which are entangled with a point in the plane (full references to the literature were given in subsection A).

Represent the configuration (self-avoiding random walk, macromolecule with excluded volume) by a set of N freely hinged links, each of length ℓ, with end points in $\vec{r}_0, \vec{r}_1, \ldots, \vec{r}_N$. All $\vec{r}_j$ are restricted to a plane. The sum over all self-avoiding configurations which start at $\vec{r}_0$ and end at $\vec{r}_N$ is given by the multiple integral

$$Q(\vec{r}_N,N|\vec{r}_0,0) = \underbrace{\int d^2\vec{r}_1 \int d^2\vec{r}_2 \ldots \int}_{\text{constraint}} d^2\vec{r}_{N-1} \prod_{j=0}^{N-1} (2\pi\ell)^{-1} \times$$

$$\times\ \delta(|\vec{r}_{j+1}-\vec{r}_j|-\ell), \tag{I.1}$$

where the 'constraint' indicates the requirement that the polygon which passes through $\vec{r}_0,\vec{r}_1,\ldots,\vec{r}_N$ will have no self-intersections. The integral (I.1) immediately exemplifies a profound difference between the two- and three-dimensional cases of this problem: in order to get non-Gaussian behavior in three dimensions one has to imagine a small hard sphere of diameter $d>0$ around each of the $\vec{r}_j$. In the limit $d\to 0$ the set of configurations with self-intersections becomes a set of measure zero and Q becomes Gaussian. In contrast with this behavior the two-dimensional problem will produce non-Gaussian statistics even in the limit $d\to 0$ because self-intersecting polygons in the plane form a finite fraction of all polygons. Hence the two-dimensional problem differs essentially from the three-dimensional one.

Denote by A a configuration $(\vec{r}_0,\vec{r}_1,\ldots,\vec{r}_N)$ without self-intersections, by B a configuration with at least one self-intersection and by C an arbitrary configuration. Suppose one could find a complex weight function W(C) with the three following properties:

1. W(A) = 1.
2. W(B) is arbitrary, but such that

$$\int d(B)\ W(B) \prod_{j=0}^{N-1} (2\pi\ell)^{-1}\ \ (|\vec{r}_{j+1}-\vec{r}_j|-\ell) = 0. \tag{I.2}$$

   Here we used a somewhat hybrid notation, in which the integration over all self-intersecting polygons with end points fixed at $\vec{r}_0$ and $\vec{r}_N$ has been denoted by $\int d(B)$.
3. W(C) is a product of N factors such that the j-th factor only depends on the position and orientation of the j-th link and its immediate predecessor.

If a complex weight with these three properties exists one can write

$$Q(\vec{r}_N,N|\vec{r}_0,0) = \int d(C)\ W(C) \prod_{j=0}^{N-1} (2\pi\ell)^{-1} \times$$

$$\times\ \delta(|\vec{r}_{j+1}-\vec{r}_j|-\ell), \tag{I.3}$$

and the evaluation of the multiple integral leads to a Markovian problem because of property 3. We shall now construct a complex weight function with the desired properties.

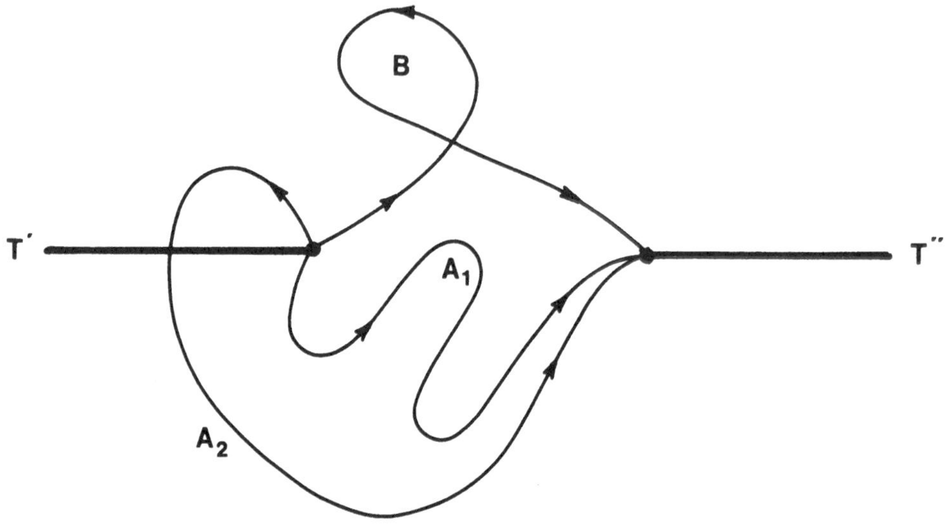

Fig. 4

Three typical configurations of a polymer chain which is constrained to a plane (discussion in subsection I).

Let $\phi_j(-\pi < \phi_j < +\pi)$ denote the increase in the angle which the tangent to the polygon makes with the straight line connecting $\vec{r}_0$ with $\vec{r}_N$ when the polygon passes through the point $\vec{r}_j$. In Fig. 4 the end points $\vec{r}_0$ and $\vec{r}_N$ have been drawn, and three configurations connecting them. It is convenient to use Cartesian coordinates (x,y) with $\vec{r}_0 = (0,0)$ and $\vec{r}_N = (R,0)$. Two lines $T' = (x,0)$ with $x < 0$ and $T'' = (x,0)$ with $x > R$ have also been drawn in this figure: their union will be denoted by T. For the proper definition of ϕ_0 we assume that a configuration C comes in along T', continues along C and leaves along T". With these provisions the total accumulated increase in the angle which the tangent to C makes with the positive x-axis is uniquely defined and equals

$$\phi(C) = \sum_{j=0}^{N} \phi_j, \qquad (-\infty < \phi(C) < +\infty). \tag{I.4}$$

Let moreover n(C) = 0,1,2... denote the number of times that C crosses T. Then the complex weight

$$W(C) = W_1(C)W_2(C), \tag{I.5}$$

$$W_1(C) = \exp\{\tfrac{i}{2}\,\phi(C)\}, \tag{I.6}$$

$$W_2(C) = (-1)^{n(C)}, \tag{I.7}$$

has the three desired properties 1-3. A proof of this can be found in a paper by the author (Wiegel, 1979e); its details will not be reproduced here.

Let $q_N(x,y,\phi)$, with $-\infty < \phi < +\infty$, denote the sum of the complex weights W(C) over all those configurations which (1) consist of N steps, (2) reach the point (x,y) at the end of the N-th step and (3) enter (x,y) with a total accumulated phase equal to ϕ. This function is of the form

$$q_N(x,y,\phi) = p_N(x,y,\phi)\,\exp\,(\tfrac{i}{2}\phi), \tag{I.8}$$

where the functions p_N take real values and are recurrently defined by the integral relation

$$p_N(x,y,\phi) = S(x-\ell\cos\phi,\ y-\ell\sin\phi,\phi)\,\cdot$$

$$\cdot\frac{1}{2\pi}\int_{-\pi}^{+\pi} p_{N-1}(x-\ell\cos\phi,\ y-\ell\sin\phi,\ \phi-\alpha)\,d\alpha, \quad (-\infty < \phi < +\infty), \tag{I.9}$$

where $S(x-\ell\cos\phi,\ y-\ell\sin\phi,\phi) = -1$ if the line connecting the point $(x-\ell\cos\phi,\ y-\ell\sin\phi)$ with the point (x,y) crosses T, and $S(x-\ell\cos\phi,\ y-\ell\sin\phi,\ \phi) = +1$ otherwise. Obviously $p_0(x,y,\phi) = \delta(x)\,\delta(y)\,\delta(\phi)$.

In terms of the eigenvalues λ_s and orthonormal eigenfunctions f_s of the integral equation

$$S(x-\ell \cos \phi, y-\ell \sin \phi, \phi) \frac{1}{2\pi} \int_{-\pi}^{+\pi} f_s(x-\ell \cos \phi, y-\ell \sin \phi, \phi-\alpha) d\alpha =$$

$$= \lambda_s f_s(x,y,\phi) \qquad (-\infty < \phi < +\infty) \qquad (I.10)$$

the solution is given by the expansion

$$p_N(x,y,\phi) = \sum_s \lambda_s^N f_s(x,y,\phi) f_s^*(0,0,0). \qquad (I.11)$$

Once the integral equation has been solved the number of self-avoiding random walks can be calculated from

$$Q(\vec{r}_N,N|\vec{r}_0,0) = \sum_{n=-\infty}^{+\infty} (-1)^n . \int_{2\pi n-\pi}^{2\pi n+\pi} p_N(R,0,\phi)\, d\phi. \qquad (I.12)$$

The last three equations once more demonstrate the Markovian nature of this problem.

It should be remarked that it will only be permitted to treat the configurations as continuous curves - as was done in this subsection - if the loops and segments between successive intersections always consist of a large number of links. This will be the case asymptotically for $N \gg 1$.

In most of the current computer enumerations, as reviewed by McKenzie (1976), a discrete model is employed in which the configurations are restricted to the bonds and vertices of some lattice. However, as the non-Gaussian features of the statistics are believed to be universal, the continuum model studied here should give results which are qualitatively identical to those of the discrete models.

It is easy to show that the method of this subsection can be generalized by some minor modifications to self-avoiding random walks in the plane in the presence of an external potential, or to such walks on a curved surface.

It is perhaps appropriate to end this subsection with the conjecture that the two-dimensional self-avoiding random-walk problem can be solved analytically in the large N limit. Yet the innocent appearance of the integral equation (I.10) is misleading and a solution of this equation might still be a formidable task.

J. On a remarkable class of two-dimensional random walks. In this subsection we study for their own sake the properties of random walks in the plane which have a weight which is related (but not identical) to the weight used in subsection I

$$W(C) = W_0(C)W_2(C). \qquad (J.1)$$

Here $W_0(C)$ denotes the a priori weight of a configuration C in the standard case of free random walks in the plane, and $W_2(C)$ is defined by (I.7). As W(C) can be negative the usual probabilistic interpretation does not apply. In this subsection we take for T the collection of points (x,0) with $x < 0$, often called "the branchline". The random walks to be discussed in this subsection are restricted in two respects as compared to the random walks of subsection I: (1) the walks have real rather than complex weights; (2) the sign of the weight W(C) is determined by the number n(C) of times that C crosses a single branchline rather than a set of two branchlines.

Let p(x,y,N) denote the sum of the real weights (J.1) over all configurations which: (1) consist of N steps; (2) start at (x_0,y_0); (3) reach (x,y) at the end of the N-th step. These functions take real values and are recurrently defined by

$$p(x,y,0) = \delta(x-x_0)\delta(y-y_0), \tag{J.2}$$

$$p(x,y,N) = \frac{1}{2\pi}\int_{-\pi}^{+\pi} S(x-\ell\cos\alpha,\ y-\ell\sin\alpha,\ \alpha)\cdot$$

$$\cdot p(x-\ell\cos\alpha,\ y-\ell\sin\alpha,\ N-1)\,d\alpha, \tag{J.3}$$

where S is defined as in subsection I.

Let B denote those points in the plane whose shortest distance to T is $\leq\ell$. If $\vec{r}\notin B$ the integral relation (J.3) simplifies to

$$p(x,y,N) = \frac{1}{2\pi}\int_{-\pi}^{+\pi} p(x-\ell\cos\alpha,\ y-\ell\sin\alpha,\ N-1)\,d\alpha. \tag{J.4}$$

If p(x,y,N) is a slowly varying function of x,y and N the integrand can be expanded in a Taylor series and the integral relation can be replaced by a differential equation

$$\frac{\partial p}{\partial N} = \frac{\ell^2}{4}\left(\frac{\partial^2 p}{\partial x^2} + \frac{\partial^2 p}{\partial y^2}\right), \tag{J.5}$$

$$\lim_{N\downarrow 0} p(x,y,N) = \delta(x-x_0)\delta(y-y_0). \tag{J.6}$$

Now consider the integral relation (J.3) in the case in which $\vec{r}\in B$ is part of the branchline T. In this case, because of the definition of S, the integral relation gives

$$\lim_{y\downarrow 0} p(x,y,N) = -\lim_{y\uparrow 0} p(x,y,N), \qquad (x<0). \tag{J.7}$$

Moreover, if the walks start on the x-axis, then every walk from $(x_0,0)$ to an arbitrary point (x,y) leads - after reflection in the x-axis - to a mirror image which leads from $(x_0,0)$ to $(x,-y)$ and which has exactly the same number of intersections with the branchline. Hence

$$p(x,y,N) = p(x,-y,N), \qquad (y_0 = 0). \tag{J.8}$$

Combination of the last two equations gives the boundary condition

$$p(x,0,N) = 0, \qquad (x < 0,\ y_0 = 0). \tag{J.9}$$

A peculiar consequence of this boundary condition arises in the case $x = 0$. According to (J.9) one has

$$\lim_{x\uparrow 0} p(x,0,N) = 0, \qquad (y_0 = 0). \tag{J.10}$$

But according to (J.3) one also has

$$\lim_{x\downarrow 0} p(x,0,N) = \frac{1}{2\pi}\int_{-\pi}^{+\pi} p(x-\ell\cos\alpha,\ y-\ell\sin\alpha,\ N-1)\,d\alpha \neq 0. \tag{J.11}$$

Hence the end point (0,0) of T is a point in which these functions have a finite discontinuity; this jump is a consequence of the geometric definition of the function S.

We can now calculate the function $p(x,y,N)$. We shall use polar coordinates (r,θ) with $-\pi < \theta < +\pi$ and $0 < r < \infty$; the initial point with Cartesian coordinates $(x_0,y_0 = 0)$ has polar coordinates $(r_0,\theta_0 = 0)$. From the definitions of $p(r,\theta,N)$ and the functions $Q_n(r,\theta,N)$ which were introduced in subsection B it can be shown that the following relation holds

$$p(r,\theta,N) = \sum_{n=-\infty}^{+\infty} Q_n(r,\theta+2\pi n,N)(-1)^n. \tag{J.12}$$

Substitution of the explicit result (B.31) for the entanglement probabilities gives

$$p(r,\theta,N) = (\pi N\ell^2)^{-1} \exp\left(-\frac{r_0^2+r^2}{N\ell^2}\right) \int_{-\infty}^{+\infty} I_{|k|}\left(\frac{2r_0r}{N\ell^2}\right) \cdot$$

$$\cdot \sum_{n=-\infty}^{+\infty} (-1)^n \exp(2\pi nki + \theta ki). \tag{J.13}$$

The summation over n can be performed explicitly with the use of the relation

$$\sum_{n=-\infty}^{\infty} \exp(2\pi nki + \pi ni) = \sum_{s=-\infty}^{+\infty} \delta(k+\tfrac{1}{2}-s). \tag{J.14}$$

This enables one to perform the integration over k in (J.13) and to find

$$p(r,\theta,N) = (\pi N\ell^2)^{-1} \exp\left(-\frac{r_0^2+r^2}{N\ell^2}\right) \sum_{s=-\infty}^{+\infty} I_{|s+\frac{1}{2}|}\left(\frac{2r_0 r}{N\ell^2}\right)\cdot$$

$$\cdot \exp\{(s+\tfrac{1}{2})\theta i\}. \qquad (J.15)$$

This formula determines $p(r,\theta,N)$ everywhere in the plane, apart from very close to the origin of coordinates. Note that (J.15) can be written in the alternative way

$$p(r,\theta,N) = \frac{2}{\pi N\ell^2} \exp\left(-\frac{r_0^2+r^2}{N\ell^2}\right) \sum_{s=0}^{\infty} I_{s+\frac{1}{2}}\left(\frac{2r_0 r}{N\ell^2}\right) \cos (s+\tfrac{1}{2})\theta. \qquad (J.16)$$

The value of $p(r,0,N)$ in the limit $r \downarrow 0$ can now be approximated by combination of (J.16) and (J.11)

$$p(0,0,N) \cong \frac{2}{\pi^2(N-1)\ell^2} \exp\left\{-\frac{r_0^2+\ell^2}{(N-1)\ell^2}\right\} \sum_{s=0}^{\infty} \frac{(-1)^s}{(s+\frac{1}{2})} I_{s+\frac{1}{2}}\left(\frac{2r_0}{(N-1)\ell}\right). \qquad (J.17)$$

The last two formulae determine the statistics of the random walks in the case of a single branchline. Further generalizations of these results, along the lines indicated in the beginning of this subsection, form the subject of current research.

BIBLIOGRAPHY

Abramowitz M. and Stegun I.A. (1970) "Handbook of Mathematical Functions", Dover, New York.

Alexander-Katz R. and Edwards S.F. (1972) J. Phys. A5, 674.

Barber M.N. and Ninham B.W. (1970) "Random and Restricted Walks", Gordon and Breach.

Bell G.I., Perelson A.S. and Pimbley G.H. Jr. (1978) "Theoretical Immunology", Marcel Dekker, New York.

Berg H.C. and Purcell E.M. (1977) Biophys. J. 20, 193.

Bretscher M.S. (1976) Nature 260, 21.

Brinkman H.C. (1947a) Physica 13, 447.

Brinkman H.C. (1947b) Proc. Acad. Sci. Amsterdam 50, 618.

Brinkman H.C. (1949a) Appl. Sci. Res. A1, 27.

Brinkman H.C. (1949b) Appl. Sci. Res. A1, 81.

Buas M. (1977) "A theoretical study of membrane diffusion and lymphocyte patching" Ph.D. Thesis, University of Maryland.

Burgers J.M. (1938) Proc. Acad. Sc. Amsterdam 16, 128.

Burgoyne P.N. (1963) J. Math. Phys. 4, 1320.

Chandrasekhar S. (1943) Rev. Mod. Phys. 15, 1.

Cherry R.J., Bürkli A., Busslinger M., Schneider G. and Parish G.R. (1976) Nature 263, 389.

Cone R.A. (1972) Nature New Biol. 236, 39.

Darcy H. (1856) "Les fontaines publiques de la ville de Dijon", Dalmont, Paris. Reference quoted in Felderhof and Deutch (1975) and in Muskat (1937).

Debye P. (1947) Phys. Rev. 71, 486.

Debye P. and Bueche A.M. (1948) J. Chem. Phys. 16, 573.

De Gennes P.G. (1969) Rep. Prog. Phys. 32, 187.

De Gennes P.G. (1972) Phys. Lett. 38A, 339.

DeLisi C. (1976) "Antigen-Antibody Interactions", Lecture Notes in Biomathematics 8, Springer, Heidelberg.

DeLisi C. and Perelson A.S. (1976). J. Theor. Biol. 62, 159.

Des Cloizeaux J. (1974) Phys. Rev. A10, 1665.

Des Cloizeaux J. (1975) J. de Phys. 36, 281.

Deutch J.M. and Felderhof B.U. (1975) J. Chem. Phys. 62, 2398.

Edwards S.F. (1965a) Proc. Phys. Soc. London 85, 613.

Edwards S.F.(1965b) in "Critical Phenomena", N.B.S. Misc. Publ. 273, 225.

Edwards S.F. (1967) Proc. Phys. Soc. London 91, 513.

Edwards S.F. (1978) in "Path Integrals", G.J. Papadopoulos and J.T. Devreese eds., Plenum Press, pg. 285.

Einstein A. (1905) Ann. der Phys. 17, 549.

Einstein A. (1906) Ann. der Phys. 19, 289.

Einstein A. (1911) Ann. der Phys. 34, 591.

Einstein A. (1956) "Investigations on the theory of the brownian motion", Dover, New York.

Emery V.J. (1975) Phys. Rev. B11, 239.

Fahey P.F., Koppel D.E., Barak L.S., Wolf D.E., Elson E.L. and Webb W.W. (1977) Science 195, 305.

Fahey P.F. and Webb W.W. (1978) Biochemistry 17, 3046.

Felderhof B.U. (1975a) Physica 80A, 63.

Felderhof B.U. (1975b) Physica 80A, 172.

Felderhof B.U. (1976a) Physica 82A, 596.

Felderhof B.U. (1976b) Physica 82A, 611.

Felderhof B.U. and Deutch J.M. (1975) J. Chem. Phys. 62, 2391.

Felderhof B.U. and Jones R.B. (1978) Physica 93A, 457.

Fowler V. and Branton D. (1977) Nature 268, 23.

Gradshteyn I.S. and Ryzhik I.M. (1965) "Tables of Integrals, Series and Products", Academic Press, London.

Hermans J.J. (1953) "Flow Properties of Disperse Systems", pg. 165-183, North Holland, Amsterdam.

Hilhorst H.J. (1976) Phys. Lett. 56A, 153.

Huang H.W. (1973) J. Theor. Biol. 40, 11.

Jones R.B. (1978a) Physica 92A, 545.

Jones R.B. (1978b) Physica 92A, 557.

Jones R.B. (1978c) Physica 92A, 571.

Jones R.B. (1979) Physica 97A, 113.

Jones R.B., Felderhof B.U. and Deutch J.M. (1975) Macromol. 8, 680.

Kac M. and Ward J.C. (1952) Phys. Rev. 88, 1332.

Kirkwood J.G. and Riseman J. (1948) J. Chem. Phys. 16, 565.

Kramers H.A. (1946) J. Chem. Phys. 14, 415.

Landau L.D. and Lifshitz E.M. (1959) "Fluid Mechanics", Addison-Wesley, Reading, Mass.

Lehninger A.L. (1972) "Biochemistry", Worth.

Lifshitz I.M., Grosberg A.Y. and Khokhlov A.R. (1978) Rev. Mod. Phys. 50, 683.

Maxwell J.C. (1856) Trans. Cambridge Philos. Soc. 10.

McCammon J.A., Deutch J.M. and Felderhof B.U. (1975) Biopol. 14, 2613.

McKenzie D.S. (1976) Phys. Rep. 27, 35.

Mijnlieff P.F. and Jaspers W.J.M. (1971) Trans. Far. Soc. 67, 1837.

Mijnlieff P.F., Jaspers W.J.M., Ooms G. and Beckers H.L. (1970) Disc. Farad. Soc. 49, 283.

Mijnlieff P.F. and Wiegel F.W. (1978) J. Pol. Sc. Physics Ed. 16, 245.

Muskat M. (1937) "The flow of homogeneous fluids through porous media", New York. Reference quoted in Brinkman (1949a).

Ooms G., Mijnlieff P.F. and Beckers H.L. (1970) J. Chem. Phys. 53, 4123.

Perelson A.S. (1979) "Mathematical Immunology"; Chapter 5 in: "Mathematical Models in Molecular and Cellular Biology", L.A. Segel ed., Cambridge Univ. Press.

Perelson A.S. and DeLisi C. (1975) J. Chem. Phys. 62, 4053.

Perelson A.S. and Wiegel F.W. (1979), in preparation.

Peterson J.M. and Fixman M. (1963) J. Chem. Phys. 39, 2516.

Poo M. and Cone R.A. (1974) Nature 247, 438.

Prager S. and Frisch H.L. (1967) J. Chem. Phys. 46, 1475.

Prives H. and Shinitzky M. (1977) Nature 268, 761.

Purcell E.M. (1977) Am. J. Phys. 45, 3.

Reuland P., Felderhof B.U. and Jones R.B. (1978) Physica 93A, 465.

Saffman P.G. (1976) J. Fluid. Mech. 73, 593.

Saffman P.G. and Delbrück M. (1975) Proc. Nat. Acad. Sc. USA 72, 3111.

Saito N. and Chen Y. (1973) J. Chem. Phys. 59, 3701.

Schlessinger J., Koppel D.E., Axelrod D., Jacobson K., Webb W.W. and Elson E.L. (1976) Proc. Nat. Acad. Sc. USA 73, 2409.

Schlessinger J., Barak L.S., Hammes G.G., Yamada K.M., Pastan I., Webb W.W. and Elson E.L. (1977) Proc. Nat. Acad. Sc. USA 74, 2909.

Schreiner G.F. and Unanue E.R. (1976) Adv. in Immunol. 24, 37. Academic Press.

Sherman S. (1960) J. Math. Phys. 1, 202.

Sherman S. (1963) J. Math. Phys. 4, 1213.

Singer S.J. and Nicolson G.L. (1972) Science 175, 720.

Van Dyke M. (1975) "Perturbation methods in fluid mechanics", The Parabolic Press, Stanford.

Vdovichenko N.V. (1965) Sov. Phys. - JETP 20, 477.

Wax N. (1954) "Selected papers on noise and stochastic processes, Dover, New York.

Weast R.C. (1974) "Handbook of Chemistry and Physics", CRC Press, Cleveland, pg. F-49.

Webb W.W. (1978) "Features and function of lateral motion on cell membrane revealed by fluorescence dynamics", in: "Frontiers of Biological Energetics", Vol. 2, Academic Press.

Wiegel F.W. (1969) "The cluster expansion for a macromolecule with self-interaction". Unpublished talk at the Yeshiva Conference on statistical mechanics, New York, 1 December 1969.

Wiegel F.W. (1972) Phys. Lett. 41A, 225.

Wiegel F.W. (1975a) Can. J. Phys. 53, 1148.

Wiegel F.W. (1975b) Phys. Reports 16, 57.

Wiegel F.W. (1977a) Unpublished lecture presented at the sixt Oaxtepec symposium on statistical mechanics, Oaxtepec, Mexico.

Wiegel F.W. (1977b) J. Chem. Phys. 67, 469.

Wiegel F.W. (1979a) J. Theor. Biol. 77, 189.

Wiegel F.W. (1979b) Phys. Lett. 70A, 112.

Wiegel F.W. (1979c) J. Phys. A12, 2385.

Wiegel F.W. (1979d) "Conformational Phase Transitions in a Macromolecule: Exactly Solvable Models", to appear in: "Phase Transitions and Critical Phenomena", Vol. 8, C. Domb and M.S. Green eds., Academic Press.

Wiegel F.W. (1979e) Physica 98A, 345.

Wiegel F.W. (1979f) "On a remarkable class of two-dimensional random walks" (in preparation).

Wiegel F.W. and Mijnlieff P.F. (1976) Physica 85A, 207.

Wiegel F.W. and Mijnlieff P.F. (1977a), Polymer 18, 636.

Wiegel F.W. and Mijnlieff P.F. (1977b), Physica 89A, 385.

Wiegel F.W. and Kox A.J. (1979) "Theories of Lipid Monolayers", Adv. Chem. Phys. 41, 195 - 228.

Wise M.N. (1979) Science 203, 1310.

Wolf D.E., Schlessinger J., Elson E.L., Webb W.W., Blumenthal R. and Henkart P. (1977) Biochemistry 16, 3476.

Yamakawa H. (1971) "Modern theory of polymer solutions", Harper and Row, New York.

INDEX

Lecture Notes in Physics

Vol. 91: Computing Methods in Applied Sciences and Engineering, 1977, II. Proceedings, 1977. Edited by R. Glowinski and J. L. Lions. VI, 359 pages. 1979.

Vol. 92: Nuclear Interactions. Proceedings, 1978. Edited by B. A. Robson. XXIV, 507 pages. 1979.

Vol. 93: Stochastic Behavior in Classical and Quantum Hamiltonian Systems. Proceedings, 1977. Edited by G. Casati and J. Ford. VI, 375 pages. 1979.

Vol. 94: Group Theoretical Methods in Physics. Proceedings, 1978. Edited by W. Beiglböck, A. Böhm and E. Takasugi. XIII, 540 pages. 1979.

Vol. 95: Quasi One-Dimensional Conductors I. Proceedings, 1978. Edited by S. Barišić, A. Bjeliš, J. R. Cooper and B. Leontić. X, 371 pages. 1979.

Vol. 96: Quasi One-Dimensional Conductors II. Proceedings 1978. Edited by S. Barišić, A. Bjeliš, J. R. Cooper and B. Leontić. XII, 461 pages. 1979.

Vol. 97: Hughston, Twistors and Particles. VIII, 153 pages. 1979.

Vol. 98: Nonlinear Problems in Theoretical Physics. Proceedings, 1978. Edited by A. F. Rañada. X, 216 pages. 1979.

Vol. 99: M. Drieschner, Voraussage – Wahrscheinlichkeit – Objekt. XI, 308 Seiten. 1979.

Vol. 100: Einstein Symposion Berlin. Proceedings 1979. Edited by H. Nelkowski et al. VIII, 550 pages. 1979.

Vol. 101: A. Martin-Löf, Statistical Mechanics and the Foundations of Thermodynamics. V, 120 pages. 1979.

Vol. 102: H. Hora, Nonlinear Plasma Dynamics at Laser Irradiation. VIII, 242 pages. 1979.

Vol. 103: P. A. Martin, Modèles en Mécanique Statistique des Processus Irréversibles. IV, 134 pages. 1979.

Vol. 104: Dynamical Critical Phenomena and Related Topics. Proceedings, 1979. Edited by Ch. P. Enz. XII, 390 pages. 1979.

Vol. 105: Dynamics and Instability of Fluid Interfaces. Proceedings, 1978. Edited by T. S. Sørensen. V, 315 pages. 1979.

Vol. 106: Feynman Path Integrals, Proceedings, 1978. Edited by S. Albeverio et al. XI, 451 pages. 1979.

Vol. 107: J. Kijowski, W. M. Tulczyjew, A Symplectic Framework for Field Theories. IV, 257 pages. 1979.

Vol. 108: Nuclear Physics with Electromagnetic Interactions. Proceedings, 1979. Edited by H. Arenhövel and D. Drechsel. IX, 509 pages. 1979.

Vol. 109: Physics of the Expanding Universe. Proceedings, 1978. Edited by M. Demiański. V, 210 pages. 1979.

Vol. 110: D. A. Park, Classical Dynamics and Its Quantum Analogues. VIII, 339 pages. 1979.

Vol. 111: H.-J. Schmidt, Axiomatic Characterization of Physical Geometry. V, 163 pages. 1979.

Vol. 112: Imaging Processes and Coherence in Physics. Proceedings, 1979. Edited by M. Schlenker et al. XIX, 577 pages. 1980.

Vol. 113: Recent Advances in the Quantum Theory of Polymers. Proceedings 1979. Edited by J.-M. André et al. V, 306 pages. 1980.

Vol. 114: Stellar Turbulence. Proceedings, 1979. Edited by D. F. Gray and J. L. Linsky. IX, 308 pages. 1980.

Vol. 115: Modern Trends in the Theory of Condensed Matter. Proceedings, 1979. Edited by A. Pekalski and J. A. Przystawa. IX, 597 pages. 1980.

Vol. 116: Mathematical Problems in Theoretical Physics. Proceedings, 1979. Edited by K. Osterwalder. VIII, 412 pages. 1980.

Vol. 117: Deep-Inelastic and Fusion Reactions with Heavy Ions. Proceedings, 1979. Edited by W. von Oertzen. XIII, 394 pages. 1980.

Vol. 118: Quantum Chromodynamics. Proceedings, 1979. Edited by J. L. Alonso and R. Tarrach. IX, 424 pages. 1980.

Vol. 119: Nuclear Spectroscopy. Proceedings, 1979. Edited by G. F. Bertsch and D. Kurath. VII, 250 pages. 1980.

Vol. 120: Nonlinear Evolution Equations and Dynamical Systems. Proceedings, 1979. Edited by M. Boiti, F. Pempinelli and G. Soliani. VI, 368 pages. 1980.

Vol. 121: F. W. Wiegel, Fluid Flow Through Porous Macromolecular Systems. V, 102 pages. 1980.

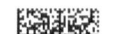